*Practical*
*Value Analysis*
*Methods*

# *Practical Value Analysis Methods*

John H. Fasal

HAYDEN BOOK COMPANY, INC., NEW YORK

ISBN 0-8104-5845-4
Library of Congress Catalog Card Number: LC 72-89911

Copyright © 1972
HAYDEN BOOK COMPANY, INC.

*Printed in the United States of America*

| 1 | 2 | 3 | 4 | 5 | 6 | 7 | 8 | 9 | PRINTING |
|---|---|---|---|---|---|---|---|---|---|

| 72 | 73 | 74 | 75 | 76 | 77 | 78 | YEAR |
|----|----|----|----|----|----|----|------|

# *Preface*

The increasingly more sophisticated methods and techniques used to apply the principles of value analysis and value engineering to ever more varied projects have made it desirable to have a reference available that deals not only with the basic concepts of traditional approaches but also with important correlated disciplines such as the theoretical concept of value, cost effectiveness, reliability control, statistics and probability, sampling, programming, decision techniques, progress control, psychological aspects of value engineering efforts, organization and implementation, and the like.

A team of recognized value analysts prepared contributions covering the different disciplines for this book, and these have been carefully interrelated in order to provide a general review of the state of the art in a science that continues to gain more and more recognition at all levels of industrial management. In the interest of providing the broadest treatment of the subject within necessary space limitations, the reader has occasionally been referred to other publications in which specific subjects have been treated in more detail. In particular, modern engineering techniques often require sophisticated mathematical approaches exceeding by far the intended level of the present volume. On the other hand, the information that is offered here will not only help the reader *choose* the best methods for his purpose but also implement and control his programs.

The editor wants to express his deep gratitude to all the contributors to this book, who will be found in alphabetical sequence on the following page.

J. H. FASAL

# List of Contributors

*Bhote, Keki*   Manager, Value Engineering and Cost Reduction, Motorola, Inc., Chicago, Ill.

*Denholm, Donald H.*   Professor of Industrial Engineering, Auburn University, Auburn, Ala. (formerly Corporate Staff member, Brown Engineering, Huntsville, Ala.)

*Fowlkes, Dusty*   President, Value Analysis, Inc., Schenectady, N.Y. (in conjunction with his staff: Eagan, Douglas L., Manager, Product Evaluation; Frick, Hank, Manager, Value Communication; Langmade, Lowell, Lecturer; Ruggles, Wayne F., Vice President, Eastern Operations)

*Garatt, Arthur*   Technical Director, Value Analysis, Inc., London, England

*Karabinos, Andrew*   Former Manager, Advanced System Engineering, Spaco, Inc., Huntsville, Alabama.

*Rasmusson, Donald*   Former Manager, Value Analysis, Walter Kidde & Co., Belleville, N. J.

# Contents

# 1
# *What Is Value?*

It is interesting and somewhat surprising to note that the key interfunctional areas of a business structure —engineering, manufacturing, marketing, finance— nearly always appear on organizational charts as blocks and not as interlocking links. Because this "block" concept is so deeply rooted, a person attached to and accountable for a specific area in a company often reacts as if he were responsible for that area alone even though he may be frequently subjected to such observations as "sales are up and profits are down", "there is a profit squeeze," or "every department must review operations and contribute to becoming more competitive and profitable."

The natural outgrowth of this block thinking is likely to be a situation similar to the following: Marketing may report that the market has been analyzed and all findings submitted to Engineering. The analysis has concluded that the product can be sold; the preliminary job has obviously been done, therefore, and now it is up to Engineering. Engineering, in turn, may state that Marketing's findings have been taken into account and the product developed to meet all performance requirements: quality, reliability, flexibility, maintainability, and appearance. The engineering job has been completed and the design passed on to Manufacturing. If the company is not competitive and the job not profitable, it is not Engineering's fault either. And so the buckpassing goes, from department to department.

Purchasing may justly claim that it has located sources and has shipped the materials into the plant on time and at competitive prices, thereby fulfilling its responsibilities. Manufacturing may take the attitude that, since the products were fabricated according to specifications and delivered to Shipping on schedule, its own responsibility has been met. Shipping, in turn, may insist that all products have been delivered with all possible dispatch. When all departments have been heard from, it seems that it is always "someone else's job" to insure profits.

Management finally recognizes that, if Purchasing cannot buy for less and Marketing cannot sell for more, only one avenue remains open: To identify and remove areas of high and unnecessary internal costs. On the surface, this seems to be the most logical step. Unfortunately, one reason many men in industry find it difficult to identify high cost areas within their own plants is that possibly they have contributed personally to the initial decisions that created them.

In studying the key interfunctional areas of a business structure, it becomes quite evident that for a company to offer its goods and services at competitive prices, a program aimed at supplementing (rather than replacing) the internal cost reduction techniques used by the key cost-creators is sorely needed. Thus, as a result of the relentless pressures of competition, the concepts of Value Analysis, Value Engineering, and Value Control were born.

# 1.1 Definitions of Value

Survey after survey has concluded that responsible men in industry cannot agree upon a one-sentence definition of "value." "Value" seems to be a word so beautifully understood without being defined that many people (and organizations) have never thought to try to explain it in terms generally acceptable to others.

To analyze the parameters of value as they apply to industry, therefore, it is necessary at the outset to focus attention on the broad concept of value.

A clear-cut definition of value can perhaps best be reached by considering the seven classes of value that Aristotle described more than two thousand years ago: (1) *Economic,* (2) *political,* (3) *social,* (4) *aesthetic,* (5) *ethical,* (6) *religious,* and (7) *judicial.*

Value analysts and value engineers are most concerned with *economic value,* which itself can be subdivided into four types: (1) *Cost value,* (2) *exchange value,* (3) *esteem value,* and (4) *use value.* These categories are not entirely separate; on the contrary, they often overlap, especially as any one or several of them may affect another. For example, esteem value and use value obviously influence exchange value strongly.

*Cost value* can be defined simply as the total cost of producing a particular item: The sum of labor, material, and overhead costs.

*Exchange value* is a measure of all the properties or qualities of an item that could make someone give something else up for it.

*Esteem value* is a measure of all the properties and features that could make its ownership desirable.

*Use value* is based on the instrumental properties or qualities of an item, the work or service it can perform or help get accomplished. This is the fundamental form of economic value. Without use value an item could have neither exchange nor esteem value.

The *real value* of a product (often referred to as "value") is a rating of the acceptance of a product by the customer and hence the final index of economic value. The higher the real value of one item over another serving the same purpose, the greater the probability of its winning the competition. The real value of a product is always relative, and it is the result of a combination of the specific value types. In general, it *increases* with higher exchange, esteem, and use value but *decreases* with higher *cost* value.

The real value of a thing depends on local and temporal conditions; an icebox has less real value in arctic than in tropical regions, and accessories for Christmas sell in December, not February.

In value analysis (VA) and value engineering (VE), needless to say, only products designed for *identical use* under *identical conditions* are compared with each other. The objective is to determine which represents the most economical choice for the specific purpose.

This task of comparison requires a crystal-clear definition of the functions the product is expected to perform and the properties necessary to make it sell. The functional abilities and properties of a product determine its performance.

## 1.2     Performance

The performance of a product can be defined as the specific amalgam of functional abilities and properties that make it suitable (and saleable) for a specific purpose. Appropriate performance requires that the product (or service) have a predetermined level of quality, reliability, interchangeability, appearance, and maintainability and that it satisfy all of the requirements of that level.

Two products may serve the same basic purpose and yet have different performance requirements, the latter being determined by the conditions under which the products are to be used. For example, a simple household thermometer and a precise laboratory thermometer serve the same purpose—the measurement of temperature. Obviously, however, the difference of their application will require a difference in their design, and this difference will be reflected in the cost value of each product and its price.

It is one of the important goals of value analysis/value engineering (VA/VE) to determine where satisfactory performance ends and excess performance begins.

If the performance of a product exceeds the requirements of its application, its real value is correspondingly lowered. For the customer, the value of a motor producing 200 horsepower in an application where only 100 horsepower is needed is less than that of a 100-hp motor, which serves the same purpose at a lower price.

If added costs result in a performance level not otherwise attainable and this level *makes the product more useful to the customer*, the expense may be justified Otherwise, added material, labor, or overhead costs are likely to increase cost value without acceptably increasing real value.

In the long run, substantial savings can be reached by eliminating unnecessary or undesired performance characteristics.

## 1.3    Comparison of Value

Value is always relative and can be established only with reference to a *value standard*. This standard is called by different names in the technical literature, and a mis-interpretation of the basic semantics involved may cause confusion and misunderstanding. It is therefore necessary to define clearly the difference between a *value standard* and *value per se*.

The value standard, henceforth to be called the *worth*, is defined as the *dollar equivalent* of the *performance* of a product. It represents the lowest cost of a part or a finished product that will reliably perform basic and secondary functions without trade-off of any of the required parameters and that has been produced using the latest materials and methods of manufacturing. The *worth* of a product therefore indicates what its performance *should cost* and serves as the basis for a comparaison with the actual costs.

The *value* of a product reflects the percentage of its worth obtainable with the materials and methods of manufacturing employed and may be determined as follows:

$$\text{Value} = \frac{\text{Worth}}{\text{Cost}}$$

One-hundred percent value is reached when the sum of the actual costs of an item equals its worth. Inappropriate materials, uneconomical methods of production, and unnecessary performance features (performance in excess of the customer's requirements) will reduce value to less than 100 percent; improved, less expensive materials, more effective techniques in manufacturing and handling, lower overhead, and elimination of unnecessary performance, on the other hand, will considerably increase the value of an item.

The *numerical* measurement of value is identical to the practical evaluation of Eq. 1.1. Costs can be expressed in terms of measurable dollars; the material costs, labor costs, and overhead allocated to the produced item are known. It is more difficult to establish the worth of a product in dollars since worth cannot be directly measured. Worth may be determined by *artful comparison*, using as data the production costs of the constituents of a product that proved to be adequate and competitive. Failing that, the actual market price of an identical competitive product can be used, but it should be corrected in consideration of possible differences in useful performance.

Another technique for determining value is based on an analysis of the functions that a specific product has to accomplish; a dollar valuation is assigned each function. These function can be evaluated intuitively (with more or less real significance) or by theoretical considerations of the direct relation between the physical aspect of a function to be performed and the costs of such performance. The function approach is today considered one of the most powerful tools in VA/VE.

## 1.4    Costs and Price

The cost value of a product has already been defined as the total of material, labor, design, and overhead costs; it should not be confused with real value and selling price. Reduction of the cost value, however, usually results in a reduction in price and therefore increases the relative value of the product for the customer, who gets the same performance for less money.

Costs are always dependent on local and temporal conditions and may vary considerably for products which are identical in performance. The higher the value of a product (lower price for the desired performance), the greater the probability of its meeting the challenge of the market.

A product may lose its value for the customer if its delivery (for cost saving or other reasons) is delayed. Delivery is therefore another extremely important parameter in determining the value of a product. Penalties for belated delivery are often imposed by the customer, who may suffer considerable financial damages if his programming is upset by such delays.

To summarize all the foregoing, a generally acceptable definition of value as employed in the VA/VE philosophy can be formulated as follows: Value is the sum of those properties embodied in a product or service that enable it to accomplish its intended functions without sacrifice of any of the parameters of high performance, low cost, and prompt delivery.

## 1.5     Value Analysis and Value Engineering

Value engineering and value analysis are organized efforts to make a product accomplish a desired function at lowest cost and thus to maximize its real value. The methods of achieving this goal are based on a systematic analysis of those factors that contribute to a satisfactory fulfillment of the functional requirements and those that do not. Value analysis and value engineering may be differentiated as follows:

*Value analysis* is the methodical investigation of all components of an exixting product with the goal of discovering and eliminating unnecessary costs without interfering with the effectiveness of the product.

*Value engineering* is a study of all possible ways of developing new products that will perform required and unequivocally defined functions at minimum cost.

Both VA and VE concentrate their cost reduction effort on a search for a new solution rather than an already accepted solution. In both disciplines trade-offs between costs and performance are not acceptable so long as the preestablished specifications and requirements are not satisfied. Neither are costs acceptable that are entailed by unnecessary performance characteristics.

The techniques applied in VA and VE are very similar and differ in only one major detail from each other: In VA the systematic analysis of a product is performed in terms of an already existing prototype of the product; all information concerning actual costs and all specifications determining required performances are obtained from past experience with the prototype itself. In VE, a preliminary solution—made at a sufficiently advanced stage of design to be considered satisfactory —has to serve as the prototype and point of departure for a value analysis. Otherwise, the same methods and techniques are applicable for both the analysis of a finished product (VA) and the analysis of a product in the design stage (VE). Basic to value analysis is unbiased and creative thinking. It involves many educational techniques and methods to guide the mind away from prejudice and the grip of habit into the direction of independent logic. Modern techniques of applied mathematics—probability, statistics, matrix analysis, and nomography, among others—are beginning to replace guesswork with precise numerical expressions to improve the reliability of design decisions.

The value of a product depends on factors partly determined by its controllable characteristics such as the quantity and types of raw materials needed, the methods and procedures of forming and machining, the performance specifications on which the design is based, and the like, and partly by circumstances and conditions outside the control of the designer such as changing market conditions and the vagaries of consumer demand. Anything that affects costs will of course affect value.

In the following paragraphs only the controllable parameters having an impact on value will be discussed. They are the only accessible starting points for a successful value analysis.

## 1.6    Analysis of Materials

Materials to be purchased for manufacturing a product fall into two principal categories: (1) "Direct" materials and (2) "indirect" materials. Each of these may be broken down into three distinct groups: (1) Raw materials, (2) parts and shelf items, and (3) subassemblies purchased in a subcontractual arrangement with a vendor.

"Direct" materials are those which are incorporated into a product and therefore contribute in a direct way to its desired performance. All raw materials for the machining of a particular part, as well as all castings, forgings, screws, and other hardware, all electrical and chemical products, all electronic components, and all subassemblies such as printed circuits or preassembled mechanical and electrical machine parts incorporated into the final product are "direct" materials.

"Indirect" materials are those which are not part

of the final product but which are necessary for its manufacture. Tools and special devices for machining and assembling components, test equipment, artwork and photographic reproduction for printed boards, chemical cleaning and etching materials, and manuals and technical literature are indirect materials. It is to be emphasized that materials that are included in the general overhead are not considered to be indirect materials.

Raw materials cannot be subjected to value analysis apart from an analysis of the production methods —such as forming and machining or chemical treatment —used to treat them. A spectacular saving in raw material costs may be completely offset by the cost increases entailed for tooling and machining or other procedures required for transforming the material into finished products. In analyzing the possibilities of cost reductions through material choice or material change, the resulting impact on the overall manufacturing costs must always be taken into consideration at the same time.

For instance, a part which has been produced by machining on a lathe could be manufactured for less money by a stamping procedure. However, this change will pay off only when large enough quantities are involved. For small quantities, the tooling cost might very well more than offset the apparent saving.

Materials selection is a complex procedure which requires the cooperation of members of the Engineering, Manufacturing, Testing, Sales and Purchasing departments. Of enormous importance is the advice of vendors, who are often in a position to propose unexpected solutions to design problems.

The analysis of raw materials may be made in connection with design of a new product or review and re-evaluation of an already existing product. When a new design is started and the general concept of the intended new product has been formulated, the first tentative listing of its functional requirements and the necessary specifications for quality level and performance also serve as the starting point for the selection and analysis of the appropriate raw materials.

A preliminary investigation for material choice will show where standard dimensions or standardized shelf items are usable and where limitations in the properties of suggested materials either have to be corrected by a change in the intended design or require acceptance of nonstandard and consequently more expensive materials. The availability of a particular facility within the manufacturing department may sometimes be an excellent reason for modifying the material choice.

In the course of the ensuing design, engineering, development, and prototype-making phases, the preselected alternative suggested materials are scrutinized, compared with each other, and evaluated. Then final specifications and tolerances are established in accordance with the practical information obtained from calculations, prototypes, and testing. Although with the release of the design from the engineering department the material problem should be solved, it often happens that late changes prove necessary; unexpected difficulties in machining, forming, or other manufacturing procedures may show up after starting the production run.

It is not possible to establish general rules for selecting and evaluating raw materials but the following basic parameters always have a decisive impact on the choice and must be taken into consideration: Functional ability, availability, producibility, reliability and costs.

*Functional ability*    The chosen material must have all the properties necessary to perform the intended function reliably, as defined by the given specifications. It must satisfy the desired requirements in performance under the operational and environmental conditions to which it will be subjected.

*Availability* The most promising material is worthless if it is not available in suffiicient quantity and sufficient time. Availability is therefore one of the most significant parameters in evaluating and selecting materials.

*Producibility* Producibility is a measure of the ease of forming and machining the raw material or—more generally—of submitting the raw material to all those operations necessary to transform it into a finished part of the final product.

In a mechanical manufacturing procedure, the most economical methods of forming and machining depend greatly on the overall shape of the specific part. It is therefore possible to establish correlations between basic shapes of finished parts and the most appropriate method or technique of producing such shapes. Many different approaches to the visualization of these basic relations have been figured out. Since almost any shape can be obtained by different methods of forming or machining, the decision as to the most appropriate method to be applied in any specific case depends on the material, the needed quantities, and the available facilities of the shop.

The forming or machining may alter the original properties of the raw material, sometimes having a favorable and sometimes an unfavorable effect on the quality of the end product, improving or handicapping its performance and functional ability. The differing effects of various methods of forming or machining must therefore be taken into account.

Producibility considerations and the requirements of performance and functional ability are obviously closely interrelated.

*Reliability*     Reliability can be defined as the degree of probability that a material will perform its intended function without failure over a stipulated period of time and under specified operational and environmental conditions.

Two products performing the same function may operate under greatly different conditions; the required reliability of the materials used in manufacturing them will therefore be different. For example, a household clock and a ship's chronometer obviously perform the same function, that of measuring time; the accuracy required of the chronometer, however, equally obviously demands a degree of materials reliability much greater than that of the household appliance. An analysis of reliability requires the simultaneous analysis of several properties and specifications, which in turn depend on the specific application of the part or of the product.

*Cost of materials*     The cost of a raw material in dollars per unit quantity is not the deciding factor in selecting materials. What counts is the piece cost of the finished (but unassembled) part, comprising the costs for raw material, labor, and direct overhead (such as tooling, machine setting, power, and the like.) The real value of a raw material increases with decreasing cost value only if producibility and performance in terms of *desirable* properties remain unchanged or are improved. Note that the word "desirable" is emphasized. If a material provides a performance far superior to what is required, a change to a less costly but still satisfactory material may result in substantial savings. On the other hand, it is sometimes possible to reduce the overall costs of a part by increasing material costs if the more expensive material permits reducing manufacturing costs.

*Production quantities*     Production quantities not

only have an effect on the purchase price of a material but also on the most economical method of forming and machining and consequently on the costs of labor and overhead.

The overall costs of a production run always have two distinct components:

1. Costs which increase in almost direct proportion to the number of pieces produced, including the material cost (plus waste) per piece, the labor cost per piece, and the direct overhead (such as power consumption and depreciation, which in turn are proportional to the machining time). All these will be called "variable costs."

2. Costs which remain invariable over the entire production run, including expenses for special tooling, machine setup, and direct overhead (such as engineering, drafting, and prototype making). This group will be referred to as "invariable costs."

In Fig. 1.1 these two cost components of a production run are represented in a coordinate system in which

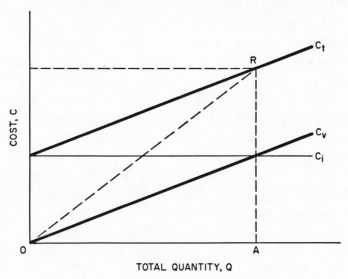

*Fig. 1.1   Variable, invariable, and total costs*

the x-axis is calibrated in terms of quantities produced and the y-axis in terms of accumulated expenses. The line $c_i$ parallel to the x-axis represents the invariable costs, and the line $c_v$ intersecting it, the variable costs. The sum of both is given by line $c_t$. Any point located on $c_t$ correlates the total costs to the produced quantity, Q.

The actual unit cost for any given quantity is given by the ratio $c_t/Q$. Thus for the given quantity of A units shown on the x-axis, the point R will correlate the total costs to this quantity, and the unit cost will be given by the ratio RA/OA.

Notice that a line passing through the origin of the coordinate system and any point on the total cost line, $c_t$, such as line OR, thus represents a specific unit cost. By means of such a line one can quickly determine the number of units one can produce for a given sum or the total costs of producing any given number of units, both in terms of any prefixed unit cost one may desire to establish.

*Changing materials in an existing product*     Material changes are always accompanied by a considerable amount of extra work, such as revision of manufacturing procedures, consideration of new tools and production devices, and drawing changes and production paperwork. Particularly is this true if material changes are made or planned for already existing products. It is therefore necessary to find out with the greatest possible exactitude if the changes will really pay. Projected savings on material costs must justify all possible resulting expenses on direct and indirect labor and overhead. If cost reduction through savings in the material cost is the only reason for the intended material change, the ratio of projected to original material costs should usually be greater than 1 : 5 to make the change profitable.

However, many reasons other than lower material

cost may justify a material change. The most important of these are: (1) Elimination of problems in production, (2) reduction of labor and other production costs, (3) improvement of the functional ability of the product, (4) increased product reliability and expected life time, (5) improvement of product appearance, (6) weight reduction, (7) taking advantage of new materials and new procedures.

Selection of new raw materials is divided into two basic steps:

1. Listing of all material properties necessary to satisfy all functional requirements.

2. Evaluating different materials to find those that provide as many of these properties as possible. As alternatives are found, a simple screening process will successively separate the best choices from the less favorable selections.

More complex cases are investigated by modern techniques, such as the following: (1) Failure analysis, using statistical approaches; (2) the theoretical evaluation of functions, which relates cost to basic functions; (3) rating and weighting systems, including the forced decision technique, which use simple matrix methods; and (4) probability calculations or graphical methods such as nomographic techniques applied to decision making.

## 1.7 Parts and Semiproducts

Small parts and components, especially if used in large quantities, may be a decisive factor in the total costs of a finished product. The right decisions in this area of semiproducts are therefore of the greatest importance for a successful value-oriented design. Despite the small and sometimes seemingly insignificant price differences between several choices, the possible savings over a long period of time may be surprising if the proper choice is made. A few simple but efficient rules in purchasing parts and components are:

1. Use standard commercial items available in standard dimensions, finishes, and tolerances.

2. Use standardized parts of identical dimension not only on one particular product but in the whole production line to get better prices for quantities purchased and to reduce storage and inventory costs.

3. Do not request narrow tolerances unless they are absolutely necessary.

4. Make sure small parts and components are available from more than one vendor, thus avoiding costly delivery delays or emergency purchasing in case of unexpected material shortage.

5. Choose vendors only after considering more than material price (location of a vendor, for instance, may reduce shipping costs and facilitate communication).

A well-organized, up-to-date filing system for literature on current components and parts is not only a great help in selecting appropriate materials and in determining standard dimensions and other specifications but supports the creative abilities of the VA team by suggesting new ideas for specific design problem solutions.

Specifications for a part or a semiproduct are in general easier to generate than for a raw material. It is often difficult to find a measurable correlation between the properties of a raw material and the requirements relating to its utilization. However, specifications for a part or component are in most cases the result of calculations made during the development of the product or obtained by practical measurements. For instance, the parameters of material fatigue, corrosion resistance, and appearance are more difficult to correlate quantitatively with properties of a specific raw material than the tolerance of a precision resistor measured in a percentage of the total resistance or the life expectancy of a reed switch defined by the number of failure-free operations of the contacts.

Before purchasing parts and components, it should be ascertained whether the design might be simplified to make the part unnecessary or to reduce its cost. This search for alternatives belongs to the creative portion of the value engineering procedure, the purpose of which is to generate ideas for a more effective, simpler, and more economical means of performing a function without sacrificing any of its required properties.

Before purchasing a part or semiproduct several specific questions should be answered. Here is a typical check list for part selection:

1. Is the part necessary for reliable performance of the end product function?

2. Is there some way to perform the function for less money?

3. Could a standard part be used?

4. Could a standard part be modified to serve the intended purpose more economically?

5. Could the design be changed to eliminate or simplify the part?

6. Could a stock item which has already been incorporated into another design be used?

7. Could the part be made more economically from another material?

8. Is the technique of forming the part the most economical?

9. Are all machined surfaces necessary or can a coarser finish be accepted?

10. Are all tolerances and specifications adequate?

11. Could the weight of the part be reduced?

12. Could screws be replaced by simpler fastening methods to speed up the assembly line?

13. Could the part be mounted with standard tools?

14. Could two or more parts be combined?

15. Are the dimensions of similar hardware parts restricted to a minimum of standard sizes?

16. Can the purchasing volume of the part be increased by using it in other areas of the design?

17. Is the part performing a basic function or is it contributing to better overall performance?

18. Is the improvement of the performance attributed to the part desirable or necessary?

## 1.8     "Make or Buy" Decision

The decision as to whether a part or semiproduct should be manufactured in the company's own facilities or purchased from a specialized and sometimes better equipped vendor is often not easily made and requires investigation and evaluation procedures very similar to those used in VA/VE. However, the general rule in deciding the "make or buy" question always remains valid: The cost of manufacturing a specific item should not exceed the price a competitor would pay for a similar part. This rule seems at first to be logical and simple, but looking closer at situations which occur frequently in practice, the problem is found to be considerably more complex and sophisticated.

One reason for the difficulty in comparing purchasing prices with manufacturing costs is that only the former are really known. Manufacturing costs may depend on many circumstances which do not yield to a direct evaluation. For instance, an idle machine that could be put into use might justify a decision to produce a part rather than buying it.

Since each piece of equipment whether producing or not is charged with a portion of the overhead, it becomes a source of loss if it is not or cannot be fully used. Such loss is reflected in reduced profit on other items manufactured in the same center. Keeping an idle machine working may make up the difference between the purchasing price and the manufacturing costs of a specific part.

Any investigation concerning the use of idle equipment should be pursued further to determine whether its reactivation will give a sufficiently high return. Because of the extremely high cost of production space, there is a strong tendency to eliminate items from the manufacturing program that give only small returns and to purchase them from a specialized vendor instead.

## 1.9 Value of Performance

The performance of a product has already been defined as the sum of all properties that make it work and sell. This definition suggests that performance has two distinct components:

1. An active component that refers to the functional ability of the product.

2. A passive component that characterizes all other properties of the product that make it desirable.

The value of performance is therefore determined by the values of the active and passive components. The value of the active component can usually be numerically calculated or assessed by artful comparison with existing alternatives. The evaluation of the passive component is more subjective and frequently depends on local and temporal considerations and on personal judgment.

The dollar equivalent for all the components of performance is called the "worth" of the performance. It forms a basis for the comparison of alternative solutions.

Remember that only those components of performance that are requested in the specifications can be taken into consideration. In technical design the active elements of performance are usually predominant so that the evaluation of performance is often identified with the evaluation of functions.

## 1.10     Value of Services

The rendering of a service can be evaluated by applying the methods and techniques used for analyzing a product. Unnecessary costs in rendering a service can be eliminated by creative thinking. A rather special service for which evaluation and cost analysis procedures are quite different from the usual ones is the consultation provided by specialists for the solution of tough problems. The profound basic knowledge of the consultant and his experience acquired after many years of intensive training determine the often extremely high fees requested. The consultation of a specialist will prove to be necessary and fruitful only if his participation in the solution of a specific problem has been well planned. The consultant must be able to concentrate his effort on the essential questions without losing time on any work that can be done by less qualified people.

The consultant's fee may be in terms of a daily charge or on a lump-sum basis. Whether the one or the other arrangement is the more desirable depends on the specific job. But in any case, the costs for consultation must constitute an acceptable portion of the development and engineering costs to which they are finally charged.

# 2

# *Functions and Costs*

The most important approach to value is the *functional approach*. It marks the difference between the usual cost reduction methods and those of value analysis. The normal cost reduction program asks the question: "How can this part be made for less money?" Value analysis asks: "How can the required *function* be performed for less money?" Keeping in mind all value concepts that relate performance and delivery to cost, VA goes one step further and asks: "How can the required function be performed in the shortest possible delivery time as well as at the lowest cost?"

## 2.1    Identification of Functions

*Function* is identified as something that makes a product work and therefore sell. It is what the product *does*. To define any required function properly is difficult. Clarity of thought is needed to distinguish a function from the specifications that prescribe the conditions within which the function must be performed. There must, of course, be practical limits to the functional approach, conforming to the ground rules within which the functions are to be defined.

For instance, if the item under study is a table lamp and the specific area to be investigated is the light bulb, the lamp itself must be considered as an accepted fact. This includes the socket into which the bulb fits and the shade that sits on top of it. Given these limits, the thinking is centered on the light bulb itself and will not be clouded by problems relating to either the socket or the shade. Consequently, the base of the bulb must also be considered as an accepted condition because otherwise it would not fit the socket.

More significant accomplishments are possible when the broad functional approach is applied to major subfunctions of systems or subassemblies of products since it is then possible to consider the relation of all parts of the overall system or design.

## 2.2    Definition of Functions

When the problem has thus been identified, two other ground rules that are important in functional definition are brought into play.

The first is to define the function in *two words*, a verb and a noun. This approach may seem arbitrary and difficult at first, but it is important for the clarity of thought that the problem requires. In selecting the two appropriate words, the value analyst is forced to think about what the object specifically does. Since every highly developed language contains many words that are not clearly understood by everyone the same way, and because the process of communication can often be difficult, the very act of defining a function in two words is almost bound to simplify understanding and communication. Electric motors "produce torque." Gas tanks "possess volume." A blast furnace "produces heat."

There will be cases where a verb phrase (such as in "bleed off air") must be substituted for a simple predicate or where an adjective will be necessary to delimit the noun so that there can be no room for ambiguity (such as in "control electric power"). It is important, however, not to go beyond three words.

The words chosen to define the function should be as broadly general as possible. Words that predetermine a single way to perform a function should be avoided. For example, if a hole is required and the function is defined as "drill hole," this definition will restrict the operation exclusively to drilling. Defining the function as "produce hole" will permit consideration of such alternatives as laser beam or electrode burning, magnetic shear, shearing punch, explosive techniques, and the like.

## 2.3     Basic and Secondary Functions

The second of the two ground rules states that there are two kinds of function: *basic* and *secondary*. A basic function is defined as the essential one for which a device is designed and manufactured. Secondary functions include all others that are subordinate to, and help support, the basic function. A simple example for basic and secondary function is that of a 35-mm camera. Its basic function, of course, is to "take picture." Among its secondary functions are such desirable features as automatic shutter release, built-in flash, interchangeable lenses, and the like. It is these secondary functions that help the product sell.

In a relatively simple assembly, there is normally only *one* basic function. As the complexity of the assembly increases, more than one basic function may be found. However, in defining two basic functions, it is necessary to make sure that there are, in fact, two functions, that the same function has not been defined in two different ways. The basic function of a chair is to "seat persons." Someone else might define it as "support weight." These two definitions are in the last analysis the same; only one basic function is involved even though a chair may support other objects besides persons. However, the definitions "provide decoration" and "seat persons" refer to two separate functions.

It has already been said that value is determined by arriving at the lowest cost of performing a function or a service reliably. Furthermore, *use value* is measured in accordance with the ability of an object to serve a specific purpose and *esteem value* on the basis of all the properties and features that could make its ownership desirable and therefore help it sell.

Producing a product at the lowest cost that will allow it both to work reliably and to sell is therefore the final goal of VA. In preparing functional definitions, emphasis has to be given to that which makes the product both work *and* sell. Keeping in mind the two ground rules of (1) defining functions in two words and (2) separating basic functions from secondary functions, one is faced by a new problem.

This problem is the tendency to confuse properties with functions. It is always important to ask the question "Is this a function that makes the product work or sell, or is it merely a property that is an adjunct of the function?" If it is a property, especially if it is a disadvantageous property, it may have to be accepted as one of the parameters within which the design philosophy is required to operate. Properties are of two types: (1) Those that are specified, such as dimensional requirements; and (2) those that are merely imposed by the overall design concept. Wherever possible, specific properties should be worked for.

To return to the light bulb mentioned previously, let us say that it has been specified to be a common 60-watt, incandescent, frosted bulb with a screw base. The specified properties of the bulb are thus at once clear, from its wattage to the type of its base (required hours of life are usually also specified). Now let us list the various things a light bulb does, without regard to whether they are to be considered as functions or properties. This list might include: it provides light, it generates heat, it attracts bugs, it aids learning, it saves eyesight, it uses electricity, it subserves decor, and so forth.

When this list is complete, or at least when no other things can be thought of, it is necessary to distinguish between the functions and the properties. Clearly all of the above items can be considered as functions with the exception of two: the attraction of bugs and the use of electricity. The latter, of course, is a specified property and the former merely a concomitant and usually a highly disadvantageous one. The next step is to evaluate the functions as either basic or secondary. After a close examination it becomes apparent that the same function may have been specified in two different ways or that a higher order of function may have been introduced.

Providing light is obviously basic because it gives the major reason why most light bulbs are designed and manufactured. Generating heat is ordinarily a mere by-product of the process required to provide light and more often than not a disadvantageous property. But, as with an infrared light bulb, it could also define a basic function. Subserving decor would ordinarily be a secondary function, although in certain instances it might conceivably be a basic one. Aiding learning should be considered a higher order of function, as should saving eyesight.

The notion of a "higher order of function" might be better understood by the following explanation: Although the basic reason for Thomas Edison's work on the light bulb was to develop a device that would replace the candle and oil and produce light with minimum fuss —not to save eyesight and to aid learning—it might have been through the latter impetus that the light bulb was finally invented.

Although the value analyst may also look for a method of producing light at lowest cost and not for a device to aid learning or save eyesight, if the definitions "aid learning" or "save eyesight" are presented as basic functions, later in the course of the analysis it will be necessary for him to search for alternative means of illumination to provide for those functions.

When its basic and secondary functions have been made clear and all its required properties specified, it becomes possible to relate the cost of the product to the function.

In most assemblies it has been found that 20 to 30 percent of the parts (or costs) perform the basic function and 70 to 80 percent of the parts perform secondary functions. This ratio of 20 to 80 percent has proved to be a most invaluable rule. The state of maturity that a product has reached can usually be determined by comparing its own basic/secondary ratio with this standard. Unnecessary costs can often be discovered, in other words, simply by separating costs into those entailed by basic and secondary functions.

Opportunities to improve value occur when a new design concept is found that allows the incorporation of needed secondary functions into the part or parts that perform basic functions. Such changes will almost surely decrease the ratio of 20 to 80 percent. The functional approach makes it possible to detect the precise areas where unnecessary costs originate.

With this objective in mind, the assembly under study (in the present case, the light bulb) must be broken down into its parts to determine which part or parts perform the basic function and which part or parts perform secondary functions. Table 2.1 shows how the breakdown into basic and secondary functions for the light bulb would be made.

**Table 2.1    Basic Functions of Incandescent
Light Bulb Parts**

| Part | Basic function of part | Does basic function of part perform basic function of entire unit? If yes, it is basic to the entire unit; if not, secondary. | |
|------|------------------------|------------------------------------------------------------------------------------|---|
| | | Basic | Secondary |
| Filament | Provide light | Yes | No |
| Glass | Enclose vacuum | No | Yes |
| Base | Provide contact and support glass | No | Yes |
| Filament support | Support filament | No | Yes |
| Sealant | Provide seal | No | Yes |
| Contact | Conduct current | No | Yes |
| Plating | Provide protection | No | Yes |
| Inert gas | Reduce oxidation | No | Yes |

The test is whether or not the individual part performs the same function as the basic function of the entire unit under study, which is here to provide light. In the given example, the filament is the only part that actually provides light. It therefore is basic, whereas all the other parts are secondary. Even if secondary parts make no contribution to the basic function of the unit, however, each must make a contribution of its own. For instance, the inert gas is not necessary to provide light, but it does reduce oxidation and thus allows the tungsten filament to burn over a longer period of time without burning up. (In "rough-service" bulbs its primary function may be to reduce fragility, in which case the function should be defined "make rugged.") If a necessary function for a specific part cannot be determined, the part should be eliminated so that the cost-function relationship will be improved.

The actual cost of each necessary part must now be thought of in terms of the basic or secondary functional performance of the part. The job of value analysis here is to try to allot as high a percentage of cost as possible to parts performing basic functions and as low a percentage of cost as possible to parts performing secondary functions. This functional approach is one of the most powerful techniques of value analysis. Even though it appears relatively straightforward, it is a technique that requires practice and skill in overcoming the problems of semantics.

## 2.4    Subdivision Into Functional Areas

Another kind of functional approach becomes a valid tool in the study of more complex assemblies. It consists in dividing such assemblies into functional *areas*. In the case of a large switch, for example, all parts performing an electrical function can be grouped together, as can all parts performing a mechanical function. Furthermore, the costs for all parts relating to the case and cover and the costs for assembly labor are computed separately, as in Table 2.2.

The engineer who sees this cost breakdown imme-

**Table 2.2    Redesigning a 10-Amp, 6-Volt Switch by Functional Areas**

| Functional area | Present cost | Cost after redesign |
|---|---|---|
| Mechanical | $10.00 | $10.00 |
| Electrical | $13.00 | $ 6.00 |
| Case and cover | $35.00 | $14.00 |
| Assembly labor | $45.00 | $35.00 |
| | $103.00 | $65.00 |

diately recognizes the obvious. Out of a total cost of $103, $23 worth of components represents the basic function of the switch. The remaining $80 was secondary expense and not at all proportional to its contribution to the amount of work performed by the switch. After an evaluation of the cost of each functional area, proper functional priorities were developed, and simplified redesign brought the cost down to that shown.

This example suggests only one classification of functional areas. Other designs will require breakdowns into other areas: Chemical, magnetic, optical, appearance, packaging, and the like. Mechanical components may be further distinguished as fixed and moving. The fixed components may then be divided into spacing, supporting, and the like. Moving components may be classified as translating, rotating, and so forth. Appearance could be broken down to include such things as machine finish, surface coating, and shape or form.

## 2.5    Functional Alternatives

The breakdown into functional areas makes it easier to search for alternative solutions. Referring again to the example of the redesigned "switch," the electrical parts proved to include gold contacts although silver contacts could do the same job and the coil supports were made of expensive machined bakelite blocks, whereas an inexpensive plastic press part was fully adequate. Likewise, the case and cover, which originally were cast, machined, and painted, were replaced by an available shelf item in finished sheet metal.

These alternative design solutions became evident because of the subdivision of the overall function of the unit into basic and secondary functions and into functional areas. Comparison of the costs of primary areas with those of secondary areas revealed unhealthily disproportionate expenditures. It should be emphasized that a systematic search for alternative solutions has to be performed for each part in a specific area because reducing the cost for *any* part contributes to the savings within that area.

## 2.6    Evaluation of Functions

A basic function can be evaluated by totaling the costs of all materials, labor, and overhead involved in its performance to the exclusion of the costs entailed by all secondary functions. These *real* costs of the function must then be compared with the costs of a standard previously analyzed. The standard may be an existing product of one's own or of competing origin, or it may have been determined by a theoretical evaluation of the required function.

The functional approach directs attention to high cost areas that are likely to induce unnecessary costs. It clearly indicates the areas where other techniques of value analysis may be applied. In summary, the functional approach is that method of thinking which differentiates value analysis from traditional cost reduction techniques.

## 2.7    Analysis of Costs

The costs of a finished product can always be attributed to three basic cost elements: (1) Materials, (2) labor, and (3) overhead.

It has been shown that the real costs of materials are determined not only by the physical properties of the raw material but also by the methods of forming and machining and by the mechanical dimensions required. These are in turn related to functional requirements, such as mechanical strength, permissible weight, size, and the like. Beyond these design considerations, the price of the material will also depend on the location of the manufacturing plant in relation to the vendor (transportation costs of raw materials are normally added to the material costs) and the quantities involved.

The costs of labor also depend in part on the methods of forming and machining that have been dictated by the overall design, with tolerances being one of the most important design parameters involved. It goes without saying that the costs of labor will vary in accordance with the location of the manufacturer, the effectiveness of the available manpower, and its proper use.

The costs of overhead are the most questionable of the three basic cost elements. They depend greatly on company policy. Overhead includes many different expenses. Some of these can be considered as more or less fixed in advance ("invariable" or true overhead), such as rent, telephone, lighting, heating, maintenance, office stationery, publicity, and salaries. Other expenses change with the type of products manufactured, such as engineering, research, and laboratory costs, depreciation of machines and equipment, and the like ("variable" overhead).

The correct distribution of overhead among the different items of a product line is vitally important for meaningful establishment of final costs and therefore a necessity for a correct value analysis of a specific item.

Overhead may be distributed in a cost analysis in many different ways. Most often it is added to labor but should be adequately allocated to all pertinent departments. In this distribution the "invariable" part of overhead is determined by the accountants. It is frequently allocated to the different departments in terms of their percentage of the budget. The "variable" part of overhead has to be calculated for each department separately and added to the "invariable" part as normal procedure.

In value analysis, however, this procedure is not necessarily the best one. The problem proves to be much more sophisticated than one might at first think.

Assume that in a specific department a product has been submitted to a successful value analysis, with the result that not only material and labor costs but also direct overhead can be reduced considerably. An equal distribution of the savings in overhead over the whole department would indeed show up as a cost saving for all items manufactured in this department but would not reflect the particular savings on the item analyzed. This inaccuracy may be of little importance to the accounting department but may very well change the decisions of management in allocating resources for the value-oriented efforts within the company. Savings in overhead as the result of a value analysis of a specific product should therefore always be included in the total saving referring to the product.

## 2.8     Cost Distribution and Cost Equations

The total costs of a product are composed of the cost of its assemblies, subassemblies, minor subassemblies, components, and the like. Each of these "semiproducts" has to be analyzed independently to recognize those areas where savings may be made in the costs of materials, labor, and overhead. Many parameters, related to each other in many different ways, are involved.

A convenient way to determine the final price of a product is to establish a *cost equation* in which all parameters affecting the total costs and their interrelations are taken into consideration. It consists of a number of additive terms related to the different types of costs (prices of the different materials including waste of material, credits for reconditioned material, setup costs, machining costs, losses from rejects, costs of overhead, and the like), which in turn always represent the sums of other similar expressions referring to the different parts of the final product.

The following equation, which is typical for a machined mechanical part, will explain the procedure better than any lengthy theoretical discussion:

$$P = \{ [\Sigma(W_m + W_m') P_m - \Sigma(W_n P_n) (1/M_m) + \Sigma(t + t_s)(P_1 + P_o)] (1 + r) + \Sigma(n_p P_p) (M_p/M_m) (1 + s) + \Sigma t'(P_1 + P_o)(1/p) + \ldots\ldots \} M_m \qquad (2.1)$$

in which:

$P$  $\;=\;$ lowest part price of produced item
$W_m$  $=$ weight of machined part
$W_m'$  $=$ weight of material waste per part
$W_n$  $=$ weight of recuperated material per part
$P_m$  $=$ lowest price for the machined material per unit weight

$P_n$ = highest price for recuperated material per unit weight

$P_1$ = lowest labor costs per unit time

$P_o$ = costs of overhead per unit time

$P_p$ = component price per piece

$n_p$ = number of components per machined part

$t$ = machining time per operation

$t'$ = manipulation and assembly time per run

$t_s$ = set up time per operation

$s$ = percentage of losses due to failures and rejects

$p$ = number of pieces produced per run

$M_m$ = mark up for production

$M_p$ = mark up for purchased components

$r$ = percentage of production rejects

The equation consists of a polynomial containing as many terms as necessary, each of which includes the sum of all other terms of similar configuration.

The terms between the brackets are multiplied by $M_m$, the production markup, and therefore represent the following cost components:

The total cost (profit included) of materials (machined, rejected, and waste) for all parts of the product is

$$(W_m + W_m')(P_m)(1 + r)(M_m)$$

Proceeds from recovered materials (production markup eliminated by multiplication with $1/M_m$) are

$$(W_n P_n)(1/M_m)(1 + r)(M_m) = (W_n P_n)(1 + r)$$

Costs for machine setups and machining per part are

$$(t + t_s)(P_1 + P_o)(1 + r)(M_m)$$

Costs of all parts bought outside and going into one piece (failures and markup on components are included) are

$$(n_p P_p)(M_p/M_m)(1 + s)(M_m) = (n_p P_p)(M_p)(1 + s)$$

Costs of the machine setup referred to the production of one piece (markup for production included) are

$$(t') \, (P_1 + P_o) \, (M_m/p)$$

The cost equation cannot be developed in a general form because its configuration depends on company policies as well as the products manufactured. The overhead, for instance, may be considered as a part of the total labor costs and be entered in the equation as an increase of hourly pay or it may be considered as part of the price of the final product. The engineering costs may appear as a separate term or may be included in the overhead. The same is true for packing, shipping, etc.

The many different ways of including nonproductive cost parameters in the overall costs of a product justify the establishment of a cost equation whenever the cost distribution for a specific and continuously manufactured product has to be recorded and analyzed. In such cases the cost equation is much more than a check list for the costs of material, labor, and overhead; it also shows the mutual relationships between the different parameters and their "weight" in determining the total cost.

It is often worthwhile to generalize the cost equation for a specific product line by incorporating all possible parameters that may be needed to evaluate any item related to the line. The necessary and unnecessary parameters of a specific product can then be separated by setting all additive terms to be eliminated equal to zero and all multiplicative terms equal to one.

## 2.9    Areas of High Costs

The discovery of areas of high costs among the many components of a product is one of the powerful tools available to the value analyst. Many techniques are available for this purpose, none of which are applied until after the total costs have been evaluated and broken down into small cost areas. The following methods are those most often used:

1. Personal judgment based on experience and comparison with similar items produced by a competitor or in a former production run at the company's own factory.
2. Determination of the cost elements (material, labor, and overhead) for all components and calculation of their proportional contribution to the overall cost.
3. Determination of the importance of the contribution to the required performance of the finished product.
4. Determination of costs per period of time.
5. Determination of costs per pound.
6. Determination of costs per dimension.
7. Determination of costs per functional property.

*Method* 1     Professional experience and common sense are often enough to discover areas of high cost in a product whose total cost has been broken down into its cost-contributing components. However, intuition and guesswork will often fail if highly sophisticated products are being investigated. In such cases, methods 2 and 3 are more appropriate.

*Method* 2     The breakdown of all costs into the three basic cost elements of material, labor, and overhead is a more precise technique for pinpointing areas of high costs. The distribution of these elements is often fairly constant for a specific production center. The figures for the normal distribution of a center can be obtained from the accounting department. If analysis of a part shows a significant deviation from this distribution, a closer look has to be taken at the cost element which causes the discrepancy.

Example: In a production center the normal cost-element distribution is 40:25:35; that is, 40 percent of the total product costs are for material, 25 percent for labor, and 35 percent for overhead. The part being investigated is a metal piece, machined in four operations on a lathe and in three operations on a drill press. The actual costs to produce the piece are $2.60 for material, $1.60 for labor, and $2.25 for overhead for a total of $6.45. Since 2.60/6.45 = 0.403, 1.60/6.45 = 0.248 and 2.25/6.45 = 0.349, the cost distribution of the part corresponds extremely well to the average distribution and is thus seen to be normal. For another part produced in the same center the cost distribution was found to be $3.60 for material, $4.00 for labor, and $5.00 for overhead, for a total of $12.60. The percentage distribution here is thus 3.60/12.60 = 0.286, 4.00/12.60 = 0.137, and 5.00/12.60 = 0.397. Since the material amounts to only 29 percent of the total price whereas labor and overhead make up more than 70 percent, it is obvious that the costs of labor and overhead are too high compared with material costs. Since the overhead is usually added to the hourly price of labor as a fixed amount that is reviewed every month, the simultaneous increase of labor and overhead in the present case was found to reflect an increase in machining time. The investigation finally showed that the additional operations required by narrower tolerances justified the deviation from the average cost-element distribution.

The example is highly instructive for another reason. It shows that the application of average cost-element distribution is not an absolute means of discovering areas of unnecessary costs but only an indicator of areas of relative high costs, which in turn may be justified or not. The ratio of labor costs to material costs may change from piece to piece if the machining time, including preparatory work, increases or if the chosen materials are not only different in price and quantity but also in machinability. In any case, the indication of high cost areas is a real help in choosing starting points for the value analysis itself.

The systematic evaluation of cost elements is very helpful for another reason. It helps to determine whether or not a specific operation may be better accomplished by a specialized and better equipped manufacturer rather than by changing one's own methods of machining, treatment of materials, and the like.

*Method 3* Another technique uses, as criterion for the permissible costs of a part, its contribution to the required performance of the finished product. In this technique it is assumed that the ratio of part cost to total product cost should correspond to the ratio relating the contribution of the part to the total performance of the final product.

The establishment of performance ratios is usually a matter of estimate rather than of precise evaluation. However, such an estimate permits the different parts of a design to be classified in terms of their importance to the final goal.

A typical example of the technique would be the analysis of a precision instrument with a movement costing only 20 percent of the total instrument price, the remaining 80 percent being allotted to the instrument housing, accessories, and the like. Obviously, this distribution represents a serious imbalance in terms of performance and calls for a cost revision of the accessory items.

A numerical evaluation of performance is possible only when performance can be expressed in terms of a specific physical quantity, for instance, speed, light intensity, weight per unit capacity, resistance to strain or stress, and the like. Nevertheless, relative performance ratings can often be obtained by applying mathematical methods, as will be seen when the forced decision technique is discussed.

*Method 4*     In high-volume production the "costs per period of time" are a good indication of the amount of savings provided by small and inexpensive parts used in large quantities. For instance, the price differences in fasteners expressed in small fractions of a cent may appear insignificant at first but may add up to considerable amounts of money over a long period of production time. Savings of this kind may sometimes be more important than those obtained on other parts as a result of a greater effort in value analysis; the latter savings may be more spectacular percentagewise but are eventually less significant because the parts are used in much smaller quantities. As a matter of fact, it is not the saving per item that counts but the saving per production period, that is, the product of saving per item times the number of items produced in a given time period. It will be shown later that the selection of small items to be used in large quantities is simplified by the application of nomograms.

*Method 5*     The application of the "cost per pound" rating is particularly indicated for the cost comparison of heavy items such as cast or welded machinery where shipping and handling costs may make the difference between several choices. For the design itself, this evaluation is less significant than that of previous techniques.

*Method 6*     The "cost per dimension" rating is a more generalized means of comparison, but it is also restricted in its area of application. Typical examples are the evaluation of tubing or electrical wiring in terms of cost per length, paint and coatings in terms of cost per square foot, and tanks or other receptacles in terms of cost per cubic foot. The technique helps locate areas of high cost when a cost comparison is made with existing similar products or with figures based on past experience.

*Method 7*    The "cost per functional property" technique—one of the more sophisticated approaches —is based on a theoretical evaluation of function. It will be shown that primary functions such as "support weight," "transmit torque," "dissipate heat," "absorb light," and the like may be evaluated by theoretical considerations that determine the quality and quantity of the material necessary to perform the required function, the most appropriate form and dimension, the most economical machining procedure, and so on. Such information helps establish a minimum price for performing the function that may serve as a basis of comparison. For instance, if a transmission has been evaluated in terms of "costs per transmitted horsepower" and the calculated figure exceeds by far the standard costs expressed in dollars per unit torque, the transmission has obviously been improperly designed.

# 3
# *The Job Plan*

The value analysis methods and techniques already discussed make it possible to select those parts, subassemblies, and assemblies that present the best prospects of obtaining maximum returns for the monies expended. Practically all constituents of a finished article are open to improvement, but improvements should be considered only if they promise high returns. Substantial savings on a component may most often be expected if it generates a large percentage of the cost of the end product or if it is a part used in large quantities.

The job plan to be discussed in this chapter has been developed through a studied consideration of the logical, systematic sequence provided by the basic techniques of Value Analysis. Subject to continuous improvement and redefinition in the past, it is today accepted as standard procedure. Despite its apparent simplicity and the logical layout of its structure, this plan is more than a predetermined work program. It is a key to unbiased thinking and an educational tool for objective judgment.

The different phases of the plan depend strongly on each other and must be followed in the correct sequence. This standard routine does not inhibit the occasional return from a later to an earlier stage of the analysis to gather additional information for the sake of a more reliable result. It is up to the analysis team to decide to what degree each phase of the total program has to be investigated.

The Job Plan is comprised of five phases:

1. Information phase
2. Definition phase (under some circumstances the definition phase may be considered part of the information phase)
3. Search or speculation phase
4. Evaluation phase
5. Execution phase.

## 3.1    The Information Phase

The information phase is part of the entire value study and the foundation on which everything else is built. It investigates the purpose of the item being analyzed and the methods and costs of realizing the item's required performance.

Two questions must be asked at the very beginning: (1) What is the item? and (2) What does it do? To answer these apparently simple questions intelligently often requires considerable research:

1. Why was the part, subassembly, or major assembly designed or produced the way it was?
2. What specifications and requirements are to be satisfied?
3. What are the limitations in size, weight, price, and so forth?

4. What are the reasons for having chosen the specified methods of forming and machining?

The most reliable sources for this kind of information are obviously the engineering and production people who were directly involved in the original design or took part in the actual production. The worker on the machine can also be an excellent source of first-hand information. Members of the sales force know the customers' requirements and specifications, the performance of competing products, and the demands of the actual market situation. The purchasing people are the best source of information on prices of materials and parts, availability, delivery, vendor capabilities, and the like.

A great deal of preparatory work is necessary to establish costs exactly. Not only must the precise amounts paid for material, parts, labor, and overhead be known, but also all information on availability, price fluctuations, shipping and storage costs, and the like. There is no limit to the amount of information that can be useful in the successful completion of a VE/VA program; even remote facts may be worth collecting and recording.

Information on negative factors may prove to be the key to the greatest improvements: (1) Why is the reject rate of some parts too high? (2) What is the reason for an apparently too-long assembly time? Such information—based on practical experience—is best obtained from the field people, or sometimes from a customer who knows the advantages and weaknesses of a product he uses.

In both cases, it is of greatest importance to separate subjective opinion from objective facts. All information must be recorded *as is*, without trying to judge it at this stage of the investigation. What is done is done, even though there may be a better way to do it. This first phase, in other words, is strictly informative. Any discussion, evaluation, or judgment of the facts is reserved for later.

## 3.2 The Definition Phase

The first (information) phase deals with the general purpose and actual cost of the object analyzed. The definition phase concentrates on the separate parts and subassemblies of the product and the primary and secondary functions they have to perform. It is the function-oriented phase, in which each part of the design has to justify itself and in which the investigator has to ask if the specific part can be simplified or modified so that it can perform the required function for less money.

The functional analysis has to concentrate with crystal clarity on the definition of the specific function the part has to perform. After listing the basic and secondary functions of each part, the investigation proceeds to separate the necessary from the unnecessary. If the finish of a product is part of the requirements—because an attractive appearance is an important factor in selling the product—an originally secondary function, "improve appearance," becomes a primary one. The broad definition, "improve appearance," permits the analysist to consider a large number of finishing methods (galvanizing, anodizing, polishing, plating, painting, and the like) and does not limit him to a specific one, which may not necessarily be the best solution.

It is already been noted that often as little as 20 percent or so of the parts and costs of a complex design are related to the primary function of the end product, whereas the remaining 80 odd percent are embodied in secondary functions. By analyzing all functions and separating them into primary and secondary, it is often possible to eliminate areas of unnecessary costs so successfully that the unfavorable ratio is reversed. Two ways of doing so are using a new design approach with a different method of performing a function or combining basic and secondary functions in one part.

A basic difference between a conventional cost reduction effort and the VA discipline is that the cost reduction effort concentrates on possible savings in materials and labor on an existing design, whereas the value analysis effort tries to obtain unchanged functional performance at lower cost by altering methods and means and by eliminating unnecessary and undesired secondary functions.

The classification of functions as basic and secondary calls automatically for subdividing the total performance into *functional areas*. For instance, parts for mechanical requirements must be separated from those for electrical specifications, and parts belonging to an optical system have to be placed in a group by themselves.

Each of these functional areas must be treated independently of all others, its having been assumed, for the time being at least, that the other areas are necessary. In this way all thinking is focused on a single part, which can then be analyzed in terms of its own basic and secondary functions. Parts not performing basic functions are classified as secondary and considered for elimination.

## 3.3    Search or Speculation Phase

"What else could do the job?" is the characteristic question that has to be answered in the course of the search phase. This is the creative portion of the analysis.

All components, subassemblies, assemblies, and other constituents of the end product are scrutinized step by step and reviewed to find out the following:

1. If the function performed by the specific part—and therefore the part itself—is necessary.

2. If the function may be accomplished by another means, working on a different principle but without sacrifice of any required design parameter.

3. If the part itself can be simplified without endangering the required performance.

Since the strength of the VA discipline depends on the cooperation of a competent team of specialized individuals, all of whom contribute from a particular point of view to a common goal, each member must have full liberty to propose solutions, which should be recorded at this stage without objections from the other members, no matter how unreal or simple-minded these solutions may appear to be. It has often happened that a proposition that appeared impractical and even ridiculous in its original form has became a very good solution after refinement and modification by other ideas. Team members must learn to free their minds from the existing, the obvious, and the conventional in order to find new ways to do a job or create new means to perform a function.

At this point, for the first time, some team members will come face-to-face with the experience of "brainstorming." Some will take to it immediately; some will resist it as an attack on their experience and intelligence; most will accept it eventually, providing it can be proved to them that it really does work. Brainstorming has its exponents and its critics, but there seems little doubt that the encouragement to hitchhike on ideas and rove in new pastures pays off with significant breakthroughs.

Why should this be? Why is a special routine needed to free the mind from its shackles? One inhibiting factor to creative thinking is undoubtedly due to formal education, which has instilled in most people an implicit belief in what may be called the "uniqueness of solution." From one's earliest days at school in mathematics and science one is taught that every problem has one, and only one, "correct" solution. This is to be found in the "Answers" at the back of the book. This principle dogs the student throughout his entire scientific or technical education and ties him down to a simplistic method of evaluation. All he need do as solutions come to mind is to look for *faults*. If there is a fault, he passes on until he finds a "faultless"—that is to say, "unique"—solution.

After graduation the student who goes into industry may meet real-life problems for the first time. These problems have a new feature: They seldom have unique solutions. Instead, there is an optimum solution, a compromise which gives the answer best fitted to the situation. But so deeply is the concept of the unique solution ingrained in him that he automatically mixes speculation and evaluation and looks for *faults* rather than *merits* in his solutions. As soon as he arrives at a solution that is feasible, he stops his creative thinking and goes on to develop this one idea, satisfied that he has the only practical answer. This habit is most difficult to combat, but at least a knowledge of a problem always helps in its elimination, and the value engineer should be aware of the inhibiting effects that a technical education can have.

In this initial stage of thinking the value analyst must not attempt to evaluate possible solutions. He has to concentrate on the basic function and forget all considerations that could divert his mind from the basic question: "What else could do the job?"

## 3.4 The Evaluation Phase

This phase is a direct continuation of the search phase. All proposed ideas and solutions are submitted to critical consideration relative to cost, feasibility, practicability, producibility, maintainability, economy, and the like. Then it must be decided whether the new solution offers a significant improvement over the existing or originally intended design.

A new way of performing a basic function may improve performance without reducing cost, or it may take over some necessary secondary functions and thus eliminate other parts or production steps that contributed to the overall costs. In both cases the proposed solution is worth further investigation.

A simple and typical example of making one part perform a *basic and secondary* function is as follows:

A relay armature had to contact a stationary contact on a relay when in its nonenergized position. This contact was originally assured by a spiral spring, acting on the movable but stiff contact arm (armature). Replacing the spring with an elastic (spring steel) contact arm satisfied the requirement of initial spring tension toward the normally closed contact, and accomplished the primary function, "provide switching", and the secondary function, "define normal position," simultaneously. The only remaining question was whether the less expensive new solution interfered with other specifications: sensitivity, adjustability, and so forth.

Costs are not always the decisive consideration if a new solution offers a significant improvement in desired performance for an acceptable increase in price. But in the first evaluation of alternatives, costs are the deciding factor. A dollar sign must be placed on each tentative solution for later comparison with existing or proposed designs. This is the first action to be taken after completing the speculation phase. Then the entire team has to weigh the pros and cons of each proposed idea, successively eliminating impractical solutions and arriving at feasible alternatives, which are further refined and from which the final choice is made.

## 3.5    The Execution Phase

We come now to the practical realization of the proposed solution. All necessary materials are investigated and delivery schedules and lowest prices determined in cooperation with prospective vendors. Information concerning expected savings in terms of quantities produced, revised specifications, improved performance, and the like is collected. Preliminary drawings for submission to management, engineering, and production are prepared. It is important that the proposed solutions be approved by engineering, production, and sales department heads; nothing is more dangerous to the successful application of a VA/VE effort than the resistance of any one of the responsible persons. The psychological aspects of the implementation of a VA/VE project require a great deal of delicacy and judgment of human nature.

The last step, with which the task of the VA team is to all intents terminated, is to prepare a report in which all facts concerning the original and the new design are submitted to management for approval. The attitude of management toward this particular VA/VE effort and any others to come depends greatly on the form in which this report is written. It must be composed in clear and persuasive language, objectively free of personal opinion. All facts must be supported by proper proofs. Exaggerated statements must be avoided.

A successful implementation of the VA/VE effort is more likely if the reports are submitted to top management rather than to a lower level, which sometimes considers a spectacular accomplishment a personal challenge.

Management is never interested in too much technical detail; it is much more interested in the ratio of return to investment. This ratio, incidentally, improves with time because it increases with the experience of the worker and his capability to do a specific job. The curve that represents the return-investment ratio as a function of time is called the *learning curve*. Interestingly, the learning curve is surprisingly consistent in various industries that manifest very different manufacturing conditions. If the returns prove to be well under the given averages, there is something wrong in the VA effort, and a review of all the facts should be considered before submitting the case to management.

# 4

# *The Scientific Approach*

The theoretical evaluation of functions is the technique that changes VA/VE from a skillful art to a science in which a relationship between cost and function can be established and in which, therefore, the function or performance is measurable in terms of a physical property. It is a method of establishing the ideal cost target for performing a specific *basic* function. Functional value is obtained by relating a required performance defined in terms of measurable parameters to the costs of the necessary materials. Only *materials* are considered in a theoretical evaluation of functions, because material costs are theoretically the only costs that affect function. Labor and overhead are investigated in the final checkout.

## 4.1    Measure of Functional Value

Since material alone performs a function and since only material costs are to be taken into consideration, theoretical functional values may be established to serve as cost targets. These can never be reached in practice, but they are very helpful in searching for alternative solutions. The functional cost target established by means of a theoretical evaluation of functions represents the ideal solution, which is attainable only at 100-percent efficiency.

The theoretical evaluation of functions is based on the establishment of a relationship between the physical parameters of the materials necessary to perform a specific function and the corresponding lowest material costs. The establishment of such a relationship will vary from case to case and can best be explained by a few typical examples.

The torsional resisting moment of a cylindrical shaft may be expressed in terms of two parameters—the diameter of the shaft and the sheer strength of its material. A second relationship, the total cost of the shaft, may be expressed in terms of its diameter, length, and density and the material price per pound. These two relationships are formulated as follows:

$$M = \pi s_t d^3 / 16 \qquad (4.1)$$
$$C = (\pi d^2 l \delta / 4)\, P \qquad (4.2)$$

in which:

$M$ = torsional resisting moment, in inch-pounds
$C$ = total material cost of the shaft, in dollars
$s_t$ = permissible torsional shear stress, in pounds per square inch
$d$ = diameter of shaft, in inches
$l$ = length of shaft, in inches
$\delta$ = density of material, in pounds per cubic inch
$P$ = material price, in dollars per pound

Solving Eq. 4.1 for d and Eq. 4.2 for C/l gives:

$$d = (16M/\pi s_t)^{1/3} \qquad (4.3)$$

and
$$C/l = (\pi d^2/4)\, \delta P \qquad (4.4)$$

Combining Eqs. 4.3 and 4.4, we have

$$C/l = (\pi/4)\, (16M/\pi s_t)^{2/3}\, \delta P \qquad (4.5)$$

This cost per length, of course, is expressed in dollars per inch.

## 4.2     Graphical Methods of Evaluation

In Eq. 4.5, $s_t$, $\delta$, and P are known for a given material and, along with the purely arithmetical factors, can be expressed by a constant, k, so that the equation can be simplified to

$$C/l = kM^{2/3} \qquad (4.6)$$

The price per unit length therefore becomes a simple function of the torque and increases with its two-thirds power.

Equation 4.6 may be written in the general form,

$$y = ax^b \qquad (4.7)$$

which, after logarithmic transformation, gives

$$\log y = \log a + b \log x \qquad (4.8)$$

This equation can be represented on log-log tracing paper as a set of parallel straight lines (lines of equal slope) which intersect the y-axis at points dictated by the constant values of "a" that define the chosen material of density, $\delta$, permissible shear stress, $s_t$, and price per pound, P. These constant values are determined by the equation,

$$a = k = (\pi/4)\,(16/\pi)^{2/3}\,(\delta P/s_t^{2/3})$$
$$= 2.81\,(\delta P/s_t^{2/3}) \tag{4.9}$$

The graphical representation of the set of curves satisfying Eq. 4.6 is the set of parallel straight lines shown in Fig. 4.1. The points $P_1$, $P_2$, and so on represent the material price per shaft length for the transmission of a given torque, T, as a function of different materials. The lowest value of P (in the given figure, $P_1$) is conventionally called the theoretical value of the function, "transmit torque."

An analysis of Eq. 4.5 shows that for a given torque the value of the function changes in proportion to a material factor, k, which includes all the needed information about a specific material. The factor k is therefore, for a given torque, the only parameter that

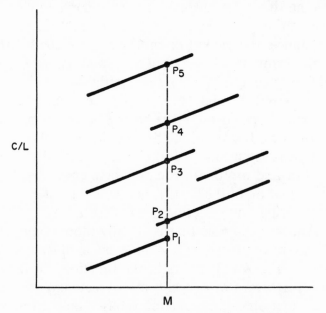

*Fig. 4.1 Graphical representation of Eq. 4.6*

determines the value of the function. For a steel shaft, it may be calculated from a table for as many alternatives as are considered useful. The lowest value determines the best choice in steel (which coincides only occasionally with the lowest material price).

Only the basic function has been considered so far in obtaining a reference for comparison. There are always many other requirements, defined by secondary functions, which have to be satisfied for the item to work and sell. Alternative ways for performing the basic function often make it possible to eliminate or combine one or more secondary functions.

In our example, the basic function, "transmit torque," is to be fulfilled by a rotating shaft. Consequently, a few secondary functions, would also have to be satisfied—"prevent rust," "reduce weight," "increase hardness," and the like. These may be taken into consideration by selecting another material than steel—for instance, the second or third choice in the list theoretically evaluated—or by making small changes in the design (replacing the steel shaft with suitable tubes of stainless steel). It might even prove necessary to modify the means of performing the basic function completely. For instance, the rotating shaft could be replaced by a pulley-belt system, a gear train, or an electrically coupled generator and motor. In each of these cases the theoretical evaluation of the basic function will be of great help in comparing and selecting the best solution.

Another example that is slightly more complicated but can be represented by an even simpler graph is shown in Fig. 4.2. It refers to the function, "resist bending," performed by a steel beam of rectangular section. The evaluation of this function is based on the following relationships:

$$M = Ws_b = Fl \qquad (4.10)$$

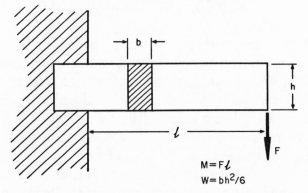

*Fig. 4.2    Beam shape versus resistance to bending*

$$W = bh^2/6 \qquad\qquad (4.11)$$

$$C = bhl\delta P \qquad\qquad (4.12)$$

in which:

M = bending momentum, in inch pounds
W = resistive momentum, in inches$^3$
C = total material price, in dollars
$s_b$ = bending strength, in pounds per square inch
F = force, in pounds
l = beam length, in inches
b = beam width (cross-sectional), in inches
h = beam height (cross-sectional), in inches
δ = density of the material, in pounds per square inch

The value equation is in this case derived as follows:

*From Eq. 4.10:*    *From Eq. 4.11:*    *From Eq. 4.12:*

$$W = M/s_b \qquad bh = 6W/h \qquad \begin{aligned} C &= 6Wl\delta P/h \\ &= 6Ml\delta P/hs_b \end{aligned}$$

Therefore, since the factor $6\delta/s_b$ is fixed for any given material,

$$C/l = (6\delta/s_b)(M/h)\,P = k\,(M/h)\,P \qquad (4.13)$$

It can be seen that in this case the material cost per length (theoretical value) depends not only on the material constant k and the unit price of the steel but also on the beam height. The theoretical value decreases with increasing beam height. This fact is easy to explain.

The resistive momentum (strength against bending) of the beam increases with the square of the beam height (Eq. 4.11), whereas the beam section and therefore its weight and total price increases only with the first power of height. Consequently, less material is necessary to resist a given bending effect with increasing beam height. To be able to make comparisons, it is therefore necessary to introduce a parameter to account for the shape of the rectangular beam. The easiest way to do so is to introduce the ratio b/h, which determines the shape of the beam section.

$$\begin{aligned} \psi &= b/h \\ h &= b/\psi \\ b &= h\psi \end{aligned} \qquad (4.14)$$

From Eq. 4.11,

$$\begin{aligned} b &= 6W/h^2 = h\psi \\ h &= (6W/\psi)^{1/3} \end{aligned}$$

Introducing Eq. 4.10,

$$h = (6M/s_b\psi)^{1/3} \qquad (4.15)$$

Introducing this value of h into Eq. 4.13 gives finally

$$\begin{aligned} C/l &= 6\delta MP/s_b h \\ &= 6\delta MP/s_b\,(6M/s_b\psi)^{1/3} \\ &= (6^{2/3}\psi^{1/3}\delta P/s_b^{2/3})\,(M)^{2/3} \qquad (4.16) \end{aligned}$$

Since $s_b$, $\psi$, $\delta$, and P are constants for a given material, this equation can be written as follows:

THE SCIENTIFIC APPROACH 61

$$C/1 = kM^{2/3} \qquad (4.17)$$

It should certainly not be surprising that for a given form of the beam section the theoretical value of the beam increases with the two-thirds power of the applied momentum.

## 4.3    Nomographic approach

The theoretical evaluation of a function is basically simple and easy to understand. As long as only a small number of parameters are involved, a more or less standard procedure can be applied. In many cases, however, the basic function depends on a large number of parameters with complex interrelationships, some of them partly of empirical character. The establishment of meaningful equations relating cost to performance then becomes extremely difficult, if not impossible.

An instructive example might be the theoretical evaluation of the function, "produce magnetic field," with a solenoid of given size. The magnetic field is determined in this case by the current flowing through the solenoid winding. The current itself is a function of the voltage applied to the coil and the coil resistance. The coil resistance depends on the coil dimensions, the number of turns, and the wire material. The number of turns is a function of the coil dimensions, the insulation, and the permissible current density, which in turn depends on the permissible coil temperature and the mode of operation (continuous or intermittent).

The cost of the coil is determined by the quantity and the size of the wire, its insulation, its form, and the like. If the coil contains an iron core, another set of correlated parameters defining the magnetic circuit has to be taken into consideration.

The most sophisticated method of conventional graphical representation would not be able to take all these parameters into account. The situation is further complicated by the fact that some of the involved variables are only empirically related to others or contain discontinuities within the usable ranges, and therefore the use of common methods of graphical representation is precluded. It is obvious that changing only one of the parameters mentioned may affect many others.

Where many variables are related to one another in such a complex way, nomographic representation becomes an extremely useful tool of the value analyst.

A nomogram in its most elementary form consists of three straight or curved calibrated scales. They are correlated by means of a straightedge, which, intersecting two scales at desired known values, crosses the third at the point whose numerical value satisfies the given relation. (See Fig. 4.3.) Straight parallel scales are the most common, but more sophisticated configurations, partly parallel and partly oblique or curved scales, are frequently necessary.

In its basic form, the nomogram correlates only three variables, but by increasing the number of scales, the number of variables can be correspondingly increased. In Fig. 4.4, for example, the straightedge is placed across six scales in four successive steps (in the first step, a straightedge is placed across known values on scales 1 and 3 and the unknown is read from scale 2). In most nomograms, the "unknown" scale lies between the two "known" scales; however, nomograms in which the desired value is on one of the outside scales, or on a scale two or more positions removed, are not rare.

The nomogram in Fig. 4.4 was prepared to determine the costs of beams of rectangular section as formulated by Eq. 4.12: $C = bhl\delta P$. Such costs can be determined with a nomogram of six logarithmic scales.

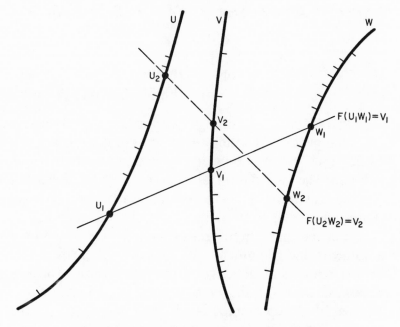

*Fig. 4.3   Example of a curved-line nomogram*

The following operations are performed, with the results of each in brackets:

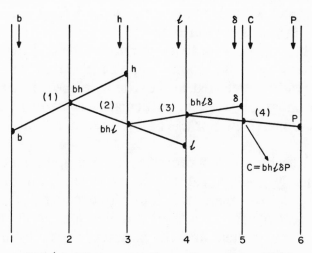

*Fig. 4.4   Nomogram: Costs of rectangular beam*

| Scale 1 | Scale 2 | Scale 3 | Scale 4 | Scale 5 | Scale 6 |
|---------|---------|---------|---------|---------|---------|
| (1) b   | [bh]    | h       |         |         |         |
| (2)     | bh      | [bhl]   | l       |         |         |
| (3)     |         | bhl     | [bhlδ]  |         |         |
| (4)     |         |         | bhlδ    | [C = bhlδP] | P   |

The sequence of nomographic operation is naturally interchangeable. It often includes complex algebraic expressions (roots, powers, exponentials, and logarithms). The nomogram also permits—by application of a special technique—the evaluation of empirical relations. This technique, called "tangent-line alignment," is one of the most important in VA/VE. The system for a tangent-line alignment (TLA) consists of two calibrated scales and an enveloping curve whose form is determined by (1) the empirical relation between the two variables represented on the two scales, and (2) the scale calibrations (linear, logarithmic, reciprocal, and so forth). If, for instance, the first scale represents the variable x in linear calibration and the second scale represents an empirical function of x, $\varepsilon(x)$, in logarithmic calibration, any tangent to the enveloping curve starting at a point $x_n$ will intersect the second (log) scale at a point $\varepsilon(x_n)$, as seen in Fig. 4.5.

It is not possible to go into further detail here about the many features the nomographic representation offers. The reader is referred to the extensive special literature on nomography (see, in particular, *Nomography*, Frederick Ungar, New York, 1968, by the author of this book).

The nomogram proves to be a graphical method eminently well suited to the representation of performance-cost relations in V/A. As a matter of fact, the requirements for an adequate representation of the involved parameters and the features offered by nomograms tend to coincide point by point, and in many

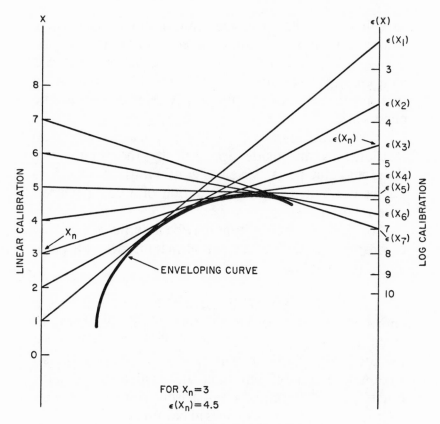

*Fig. 4.5    Tangent-line alignments*

cases justify the effort and cost of establishing the latter. The most important advantages of a nomographic representation may be summarized as follows:

1. The correlation of three or more variables is simple and accurate.

2. A change in one parameter involving changes in several other variables is easily visualized and determination of the most appropriate design solution is a matter of only a few alignments.

3. The rate of change of any parameter with respect

to others can be observed continuously; the whole system of alignments is interconnected like a set of levers.

4. Strictly mathematical relations can be combined with empirical relations and it is possible to pass from the one to the other continuously, eliminating erratic guesswork.

5. Discontinuities in either physical or empirical relationships are made apparent by the nomogram. Random value-pairs as well as continuous sets of values can be correlated for the evaluation of a cost-performance relationship.

6. The scale calibration may be in terms of a quality or property as well as the actual magnitudes of a parameter or variable.

Although the VA/VE usefulness of nomograms is enormous, it must be realized that it will not always be economical to prepare them unless they are designed for repeated use. They are particularly helpful for the selection of parts or materials, the evaluation of basic functions, and the rapid comparison of several possible solutions arising from changes in parameters. The following examples, originally developed for Value Analysis, Inc., Schenectady, are reproduced here with the kind permission of that company.

## 4.4    Examples of Nomographic Solutions

The first nomogram, presented in two parts (Figs 4.6 and 4.7), has been established for the selection of screws that are available in two different qualities of material. It is a typical example of the combination of a purely mathematical relation and an empirical relation. The former defines the screw diameter as a function of

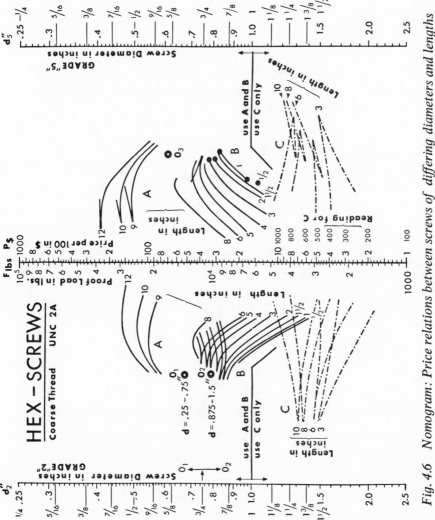

Fig. 4.6   Nomogram: Price relations between screws of differing diameters and lengths

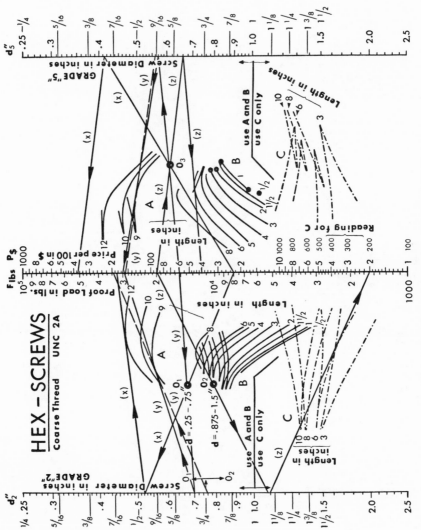

*Fig. 4.7   How alignments are carried out*

the holding force and the yield strength of the screw material:

$$d = f(F, s_t)$$

in which:

d = screw diameter
F = holding force
$s_t$ = minimum yield strength

The latter is an empirical relation between the screw diameter and screw price that has been established for different screw lengths.

The equations for the determination of the screw diameter have been derived from standard tables edited by screw manufacturers. In accordance with the expectations, it was found that the relation has a parabolic character, but that the adopted safety factor is considerably larger for grade 2 screws of large diameter than of small diameter. The discontinuity appears at the point, d = 7/8 in.

By conventional methods of reconstituting the mathematical formulation of a relation by assembling a large number of coordinated pairs of values, the following formulas have been deduced:

*Grade 2 screws:*

For 1/4″ - 3/4″ diam., $F = 32000d^{2.09}$
For 7/8″ - 1 1/2″ diam., $F = 16960d^{2.06}$

*Grade 5 screws:*

For 1/4″ - 1 1/2″ diam., $F = 45700d^{2.07}$

These relations are represented in the nomogram by scale F (common for both grades); reference points $0_1$, $0_2$, and $0_3$; and scales $d_2$ and $d_5$. The empirical relation determining the screw price per 100 pieces as a function of the screw diameter is represented by TLA's with one reference curve for each screw length.

In both representations the left side of the nomogram refers to grade 2 screws and the right side to grade 5 screws.

To cover the full range from 1/4 to 2.5 in. diameter in one graph, two sets of reference curves are provided. The first, A and B, covers the range from 1/4 to 1 in. diameter; the second, C, covers the range from 1 to 2.5 in. diameter. A discontinuity in the price-diameter relation is apparent at the overlapping point of the two ranges. Furthermore, it can be seen that there exists another discontinuity in the price relation for screws between 8 and 9 in. long, characterized by the large gap in the distribution of reference curves A and B. This discontinuity may be caused by a change in the machining procedure of screws exceeding a certain over-all size.

The use of the nomogram is very simple. In general, the starting point is the desired holding force, $F$. The point corresponding to the required proof load is located on the $F$ scale (center scale, left) and aligned with either point $0_1$ or $0_2$. Point $0_1$ is used if the alignment intersects the left $d_2$ scale in its upper part (a $d_2$ reading smaller than 0.75), and $0_2$ is used if the alignment intersects the $d_2$ scale in its lower part (a $d_2$ reading larger than 0.75). (The point 0.75 where this range change occurs is shown by a set of arrows marked $0_1$–$0_2$.) The screw diameter for grade 2 steel is read on the intersection point between the alignment and the $d_2$ scale. The screw diameter $d_5$ for the same proof load as before, but for a screw made in grade 5 steel, is found in a similar way. The alignment starting from the same point on the $F$ scale is now referred to point $0_3$, and the diameter $d_5$ is read out on the right scale, $d$.

*Price comparison*     After determination of the

diameters $d_2$ and $d_5$, which have to be rounded off to the nearest standard size, the respective prices are determined by tracing from points $d_2$ and $d_5$ tangent lines to that curve of the set of scales A, B, or C which corresponds to the desired screw length. The intersection points of these alignments with the right center scale F correspond to the respective prices.

For all diameters between 1/4 and 1 in. the curves A and B (covering the ranges 9 to 12 in. and 1/2 to 8 in., respectively) have to be used, and the F readings are made on the main calibration of the F scale (1 to 1000). For diameters between 1 and 1 1/2 in. the set of curves C has to be used, and the readings have to be made on the extended scale (100 to 1000).

Comparing the results obtained for grade 2 and 5 material, it will be found that sometimes the one, and sometimes the other, choice will be the less expensive.

In the example illustrated, three different screw sizes have been evaluated:

1. The lines x referring to a holding force of 8,000 lb and a length of 12 in.
2. The lines y referring to a holding force of 15,000 lb and a length of 9 in.
3. The lines z referring to a holding force of 20,000 lb and a length of 6 in.

It is easy to recognize that in the first case the less expensive steel (grade 2) is more economical than the more expensive steel (grade 5) and that in the last case, the more expensive steel (grade 5) is by far the better choice (40:200). In the second case there is almost no price difference between the two materials.

*Corrections*    The prices read on scale P are not to be considered as market prices but only as a means to compare costs. It has to be noted that the curves that are based on empirical parameters represent mean values and that the real prices differ somewhat from these values at distinct points (corresponding to unusual screw dimensions or to the most often used screw dimensions). The reason for the difference is the price variation brought about by manufacturing in large (1/2-in. screw) and small (9/16-in. screw) quantities. In order to take account of these natural price differences a certain percentage can be added to the price of 9/16-in. screws or subtracted from the price of 1/2-in. screws.

The nomogram in Fig. 4.8 has been established for the evaluation of a cylindrical shaft transmitting torque and provides the following information:

1. The diameter of the shaft as a function of the yield strength of the chosen steel quality and the maximum torque to be transmitted.

2. The price per unit shaft length as function of the steel price per pound and the weight per unit length.

It is a typical example for the combination of a strictly mathematical relation with empirical information consisting of random distributed pairs of correlated parameter values.

The first, strictly mathematical relation concerns the equilibrium between internal and external forces:

$$T = (\pi/16)\, d^3 S_s$$

in which:

    T  = torque
    d  = shaft diameter
    $S_s$ = minimum yield strength (shearing)

This relation is represented by the scales A, B, and C. (A for torque T, B for yield strength $S_s$, C for shaft diameter d.)

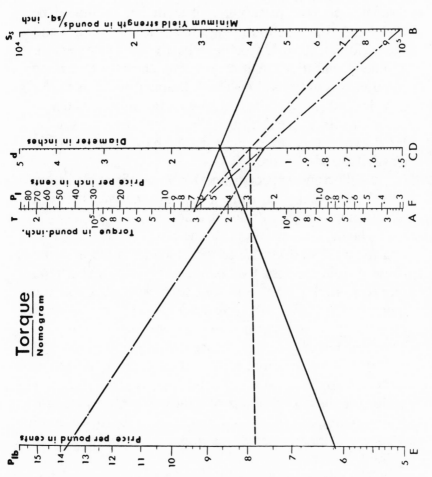

*Fig. 4.8   Torque versus price of cylindrical shaft*

The density of carbon steels is practically invariant; the weight per unit length depends only on the diameter, which in turn is a function of the yield strength of the material. Therefore, in the second (empirical) relation, the variable weight may be replaced by the variable yield strength, and the price per unit shaft length should be expressed as a function of the steel price per pound and the yield strength. Unfortunately, the establishment of a general empirical relation between strength and price per pound is not possible. This relation is noncontinuous and depends on the alloy, whose choice in turn is determined by other specifications than strength; for instance, by its resistance to oxidation, its hardness, stiffness, elasticity, and the like.

In choosing a specific steel for a specific application, there will usually be only a few alternatives between which the nomogram must decide.

The correlated values of the price per pound, $P_{1b}$, and the yield strength $S_s$, are obtained from the material chart in which these two parameters are listed for 42 differeat steel alloys. Knowing the yield strength, the diameter is found by using scales A, B, and C, and correlating scale C with scale D (representing the steel price per pound). The steel price per length can be found on scale E. In addition, a tabulating scale calibrated in terms of weight can be established opposite the diameter scale. All these results are obtained by only two alignments. Comparison of the results related to different steel qualities will give the most economical choice quickly.

The next nomogram (Fig. 4.9) is used for the determination of a rectangular bar submitted to bending. It is instructive because it is a typical example of how quickly a nomographic representation can be used to select the most convenient sets of correlated parameters by the simple displacement of alignments. Here, too,

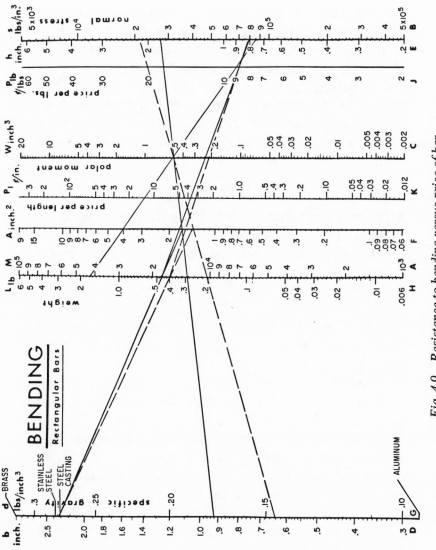

*Fig. 4.9   Resistance to bending versus price of bar*

the correlated values of the yield strength and the price per pound as function of the chosen steel quality are obtained from a steel table.

The different parameters are evaluated as follows: The theoretical bending momentum has to be determined first, and the value thus obtained has to be multiplied by a safety factor, $K_s$, to get the practical momentum on which further evaluations are based. From here on all other parameters are obtained from the nomogram.

The necessary resistive momentum of the rectangular bar is given as the ratio of bending momentum to yield strength,

$$W = M/s$$

which may be evaluated by aligning the points M, s, and W on scales A, B, and C.

On the other hand, the resistive momentum is a function of the bar width, b, and the bar section, A:

$$W = bh^{1/3} \qquad A = bh$$

The desired resistive momentum may be reached by an infinity of value pairs, b and h, which themselves correspond to an infinity of bar sections, A. Any alignment passing through point W (on scale C) correlates another set of values, b, h, and A, which satisfy the basic equations and are read on scales D, E, and F.

The most convenient dimensions of the bar have to be selected by rotating the alignment around point W and by determining sets of correlated values for the bar width, the bar height, and the bar section.

The next step deals with the evaluation of the bar weight per unit length as a function of the density, d, and the bar section, A:

$$L_1 = dA$$

The correlated values are found by aligning point

A on scale E with point d on scale G, which at the intersection with scale H determines the value of the bar weight, $L_1$, per unit length.

A last alignment gives as final result the bar price per unit length, $P_1$, on scale K as a function of the bar weight, $L_1$, on scale H and the material price per unit weight, $P_{1b}$, on scale J.

$$P_1 = P_{1b} L_1$$

In the practical example shown, the total momentum is supposed to be 42,000 lb-in. (theoretical momentum of 350 lb × 40 in., with safety factor of 3). The supporting bar is made from steel with a yield strength of 84,000 lb/sq in. and a density of 0.28 lb/cu in. A first alignment determines the necessary resistive momentum to be 0.5 in.[3]

The following calculation refers to two different solutions. The first assumes that the bar section has the dimensions, 0.92 × 1.84 = 1.7 sq in. The second solution assumes that the bar section has the dimensions, 0.65 × 2.2 = 1.4 sq. in.

The superposition of these two solutions ($P_1$ = 3.6¢/inch and $P_1$ = 3¢/inch) is a typical visual example of the rates of change to which the parameters of a relation are submitted if one parameter changes its value.

The last example (Fig. 4.10), a nomogram for the evaluation of magnetic wire, is interesting because it shows that with a relatively simple nomographic representation several pieces of information may be obtained almost instantaneously. The following data is necessary to begin with: wire length/conductivity (L/k) and current/specific load (I/σ), the latter representing the permissible load.

$$R = L/kA \qquad L/k = AR$$
$$I = \sigma A \qquad I/\sigma = A$$

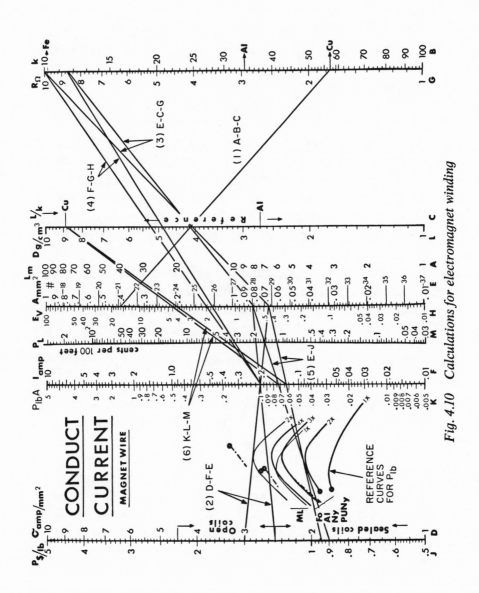

Fig. 4.10 Calculations for electromagnet winding

in which:

A = wire section, in mm²

L = wire length, in meters

k = coefficient of conductivity (k = 1/ρ in which ρ is the specific resistance in ohms per meter)

R = wire resistance, in ohms

σ = specific load permissible under the given conditions (sealed or open coil), in amp/mm²

I = current in the windings, in amps

The nomogram now determines the following:

1. The wire section as a function of both L/k and I/σ.

2. The coil resistance and the voltage drop across the coil as a function of the wire section determined in the first step:

$$E_v = IR$$

3. The wire price per pound as a function of the wire section and wire type (this is a continuous empirical function which may be represented by means of a TLA):

$$P_{1b} = F_{emp}(A)$$

in which:

P/lb      = wire price in $ per pound

$F_{emp}(A)$ = empirical function of section A

4. The wire price per 100 foot of wire as a function of the specific gravity (wire material) and the information obtained in step 3:

$$P_L = P_{1b}ALdc$$

in which:

$P_L$ = wire price per 100 ft, in cents

L   = wire length, in inches (not meters)

    d  = wire specific gravity, in grams per cu cm
    c  = conversion factor

As additional information, the scale for the specific load indicates average values of σ for different load conditions. This information has to be considered as a guide only; the correct value to be chosen in each case has to be adapted to the requirements and specifications of the designer.

The nomogram consists of two different parts: The first corresponding to steps 1 and 2, is based on strictly mathematical and physical relations. The second, comprising steps 3 and 4, is empirical and may require corrections and alterations when referring to other wire makes or to a later price schedule.

The evaluation of the different parameters is obtained as follows:

1. Given are the wire length (by the number of turns and the average winding length) on scale A and the wire material specific conductivity (k = 57 for Cu) on scale B. The ratio L/k as an auxiliary variable is read on scale C.

2. The current I on scale F and the operation conditions which determine the permissible specific load, σ, represented on scale D determine the necessary wire section, A, on scale E.

3. By aligning point A on scale E with reference point L/k on scale C, the value of the wire resistance, R, is found on scale G.

4. The voltage drop across the coil is found by aligning point I on scale F with point R on scale G and is read on scale H.

5. The calculations up to this point have a purely mathematical character. The empirical part of the nomogram comprises scales A and J and the set of reference

curves which relate the wire section, A, on scale E to the wire price per pound, $P_{lb}$, on scale J.

6. The TLA relating the two parameters to each other determines also another auxiliary variable, $P_{lb}A$, on scale K, which after alignment with the material density, d, on scale L leads finally to the desired result: the price for 100 ft of wire read on scale M.

In the example the same coil is evaluated for a current-density of $2.5A/mm^2$ and a current density of $3A/mm^2$ and the resulting changes of other parameters are clearly visualized. The evaluation is based on the following parameters:

> Wire length, L = 40 m
> Specific conductivity, k = 57 (copper)
> Max. current, I = 0.2A
> Permissible current density,
>     $\sigma$ = 2.5 or 3 $A/mm^2$
> Wire type: FO 2x
> Density, d = 0.89 lb/cu in.

# 5

# *Cost Effectiveness*

Awareness of cost effectiveness has always been an economic necessity, but it was the coining of the phrase "value engineering" by General Electric in 1947 that thrust the idea into national prominence. Adoption of the term by the Navy in 1954, and by the Army, the Air Force, and the Department of Defense before 1963, as well as growing industry acceptance, have established value engineering as a fundamental discipline in American industry.

Value engineering uses dollars to measure its effectiveness; in other words, it operates on a basis of cost effectiveness The latter has been redefined as follows: Cost effectiveness is the value engineering discipline used to analyze performance functions, from those of the individual part to those of an entire system, so as to insure maximum productivity or system effectiveness at lowest cost consistent with the restraints imposed by schedule, quantity, reliability, safety, and maintainability.

# 5.1    Worker Effectiveness Analysis

A natural starting point in cost effectiveness analysis is the determination of the effectiveness of the individual worker, but before this step can be taken it is necessary to observe and measure the effectiveness of the existing production setup and methods. It is possible that these may be improved by time and motion studies of the type originated by Frederick W. Taylor, the Gilbreths, and others.

A three-step technique can be used to analyze the activities and movements of a man at work. The record of this analysis is known as a *Man Movement Chart*.

Step 1:   Select and define symbols to represent the worker's activities such as those (adopted by the American Institute of Industrial Engineers) that are shown in Fig. 5.1(A). Record these activities on a chart—such as that in Fig. 5.1(B)—with four columns, one for describing the activity, and the other three for the appropriate symbol, the distance covered, and the time estimate.

Step 2:   Map the actual movements of the worker using an area plan such as that shown in Fig. 5.2. This plan is usually attached to the process chart for use in later analysis. (Colored pencils will help identify individual paths if more than one worker's movements are charted.)

Step 3:   Investigate the possibility of shortening and improving movement paths. Figure 5.3 shows a new layout and summarizes the savings in distance walked by the worker. No operations were changed or eliminated in this case, but if any had it would have been necessary to draw an improved chart similar to Fig. 5.2.

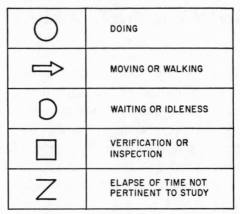

*Fig. 5.1A     AIIE symbols to denote worker activity*

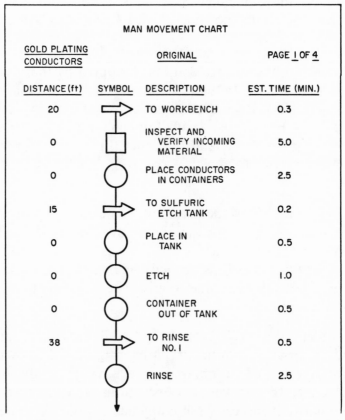

*Fig. 5.1B     Use of symbols on movement chart*

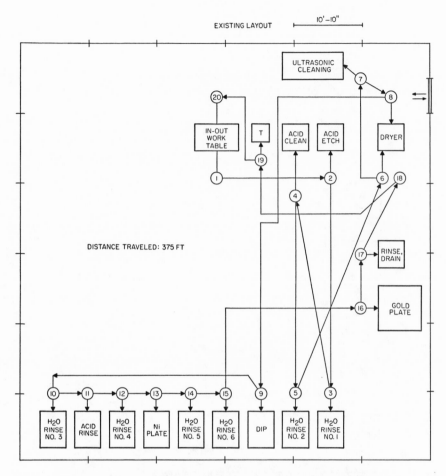

*Fig. 5.2   Man-movement chart for plating process:
(1) Arrival and inspection of parts, (2) $H_2SO_4$
etch, (3) deionized $H_2O$ rinse, (4) $HNO_3$
clean, (5) $H_2O$ (plus alcohol) rinse, (6) air-
dry centrifuge, (7) ultrasonic Freon clean, (8)
air-dry centrifuge, (9) brite dip, clean, (10)
deionized $H_2O$ rinse, (11) $H_2SO_4$ rinse, (12)
$H_2O$ rinse, (13) nickel plate, (14) $H_2O$ rinse,
(15) deionized $H_2O$ rinse, (16) gold plate,
(17) rinse, drain out, (18) air-dry centrifuge,
(19) nitric acid test, (20) inspect, log, pack,
ship*

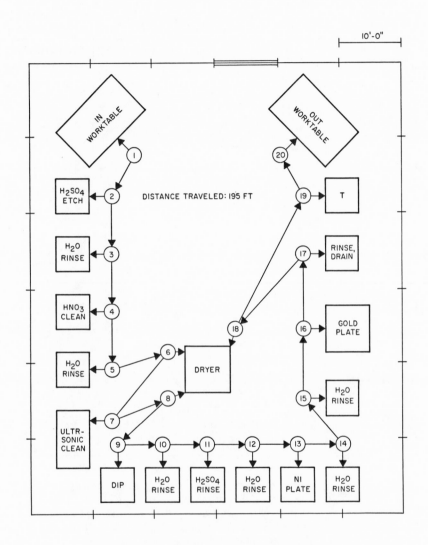

*Fig. 5.3    Revision of chart in Fig. 5.2 to reduce man-movement distances (step numbers the same)*

The *Work Place Chart* details what the operator's right and left hands do when he is working at one place. Again symbols are needed. They are similar to those of Fig. 5.1(A) except that they describe hand movements. Reference 3 at the end of the chapter gives an excellent list of such symbols. The following steps are involved in the Work place Chart:

Step 1: A sketch is traced (as in Fig. 5.4) that shows the original work place layout and describes tasks.

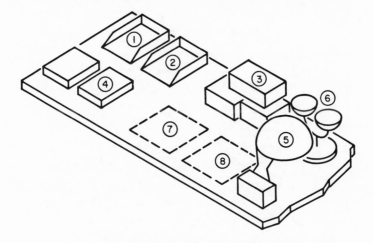

*Fig. 5.4 Work table drawing for handmotion study: (1) Incoming file, (2) outgoing file, (3) incoming work boxes, (4) outgoing work boxes, (5) shadowgraph inspection, (6) scale for sample determination, (7) area for accepted work boxes, (8) area for rejected work boxes*

Step 2: The left and right hand movements are charted with symbols (as in Fig. 5.5).

An analysis of the best layout may be supported by asking a series of pertinent questions as an aid to revision:

1. Why is the job to be done?

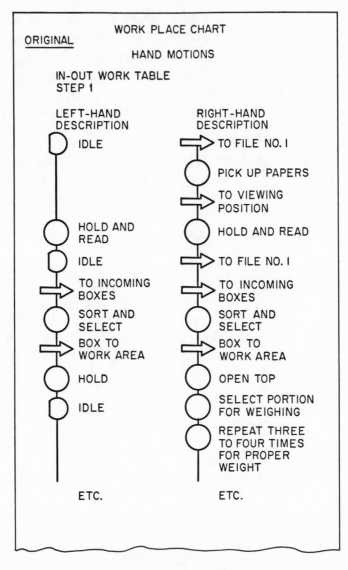

*Fig. 5.5 Symbol chart of hand motions*

2. Who is to do it?
3. Where is it to be done?
4. When is it to be done?
5. How is it to be done?

Other pertinent questions will arise, depending on the individual case.

Once the answers have been obtained, another set of questions can be asked: Can the job be eliminated? Can it be simplified? Rearranged? Combined? Again other alternatives may appear in individual cases.

As a result of the conclusions drawn in Step 2, a third step will require that a revised work place layout (Fig. 5.6) including all proposed improvements be established. Complete analysis demands that Steps 1, 2 and 3 be performed for each activity at each work place. In the case of Fig. 5.2, this means that activities at 20 work places have to be analyzed.

A *Multi-Activity Chart* shows the relationship be-

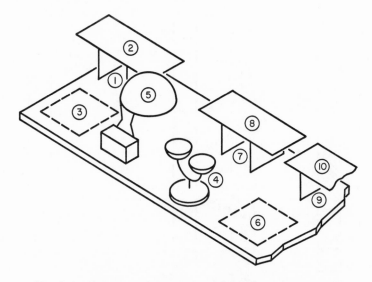

*Fig. 5.6   Revision of the work table of Fig. 5.4: (1) File basket for accepted papers, (2) shelf for accepted work boxes, (3) inspection area, (4) scale for sample determination, (5) shadow-graph inspection, (6) reassembly of lots, (7) file basket for incoming papers, (8) shelf for incoming boxes, (9) file basket for rejected papers, (10) shelf for rejected boxes*

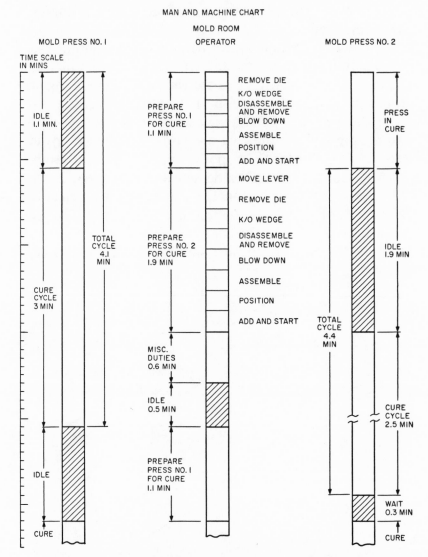

*Fig. 5.7 Time analysis for one man working two machines (Note: first machine is utilized fully, but the second has an enforced wait, with operator being idle for one-half minute per cycle; it is thus recommended that the time cycle for press 2 be shortened to match that of press 1 or that the operator be assigned additional duties)*

tween two or more workers or a man and a machine, or a man and two or more machines. It is usually the best practice to use simple bars representing a time scale as in Fig. 5.7, which charts the work of one operator tending two compression molding presses. Sometimes (as with two workers) symbols similar to those of the work place chart are used to show what each operator's hand is doing. Again, a check list helps find improved ways of accomplishing the task.

The *Therblig Chart* is used to record, without reference to work place, the fine hand motions in high-volume work. ("Therblig" is Gilbreth spelled as nearly backwards as is pronounceable.) When time periods are included in the listing of fine detail, the chart is often called a SIMO (Simultaneous Motion) chart. Because the movements are so small, it is customary to film the activity. Projecting the film one frame at a time allows the analyst to count the frames concerned with a particular activity; knowing the rate of frame exposure, the activity time can be determined with a high degree of accuracy.

The chart uses basic activity descriptions as defined by the Gilbreths. The basic Therbligs are:

| | | | |
|---|---|---|---|
| Assemble | A | Position | P |
| Disassemble | D | Pre-position | PP |
| Avoidable delay | AD | Release load | RL |
| Unavoidable delay | UD | Rest | R |
| Find | F | Search | SH |
| Grasp | G | Select | ST |
| Hold | H | Transport empty | TE |
| Inspect | I | Transport loaded | TL |
| Plan | PN | Use | U |

As an example, visualize your hand reaching for one of several pencils on a desk, in order to start writing. The motions could be charted as follows:

| Left-Hand Activities | | Right-Hand Activities | |
|---|---|---|---|
| *Therblig symbol* | *Description* | *Therblig symbol* | *Description* |
| I | Idle | TE | To pencil |
| I | Idle | ST | Pick one pencil from others |
| I | Idle | G | First contact with pencil |
| TE | Move to paper | TL | Pencil to paper area |
| H | Hold paper for right hand | P | Position pencil point to appropriate line on paper |
| H | Continue holding | U | Begin writing |

It is obvious that such an intensely detailed description would be useful as an effective cost analysis tool only for highly complex and repetitive operations.

For more detailed information on the work of the Gilbreths and others, see references 4, 5, and 6 at the end of this chapter.


## 5.2    Other Analysis Techniques


A *Paperwork Movement Chart* can be a discouraging object to prepare, but when it is completed it makes it possible to have a better understanding of "what happens to the third copy" and who has responsibility and authority for what. Figure 5.8 clearly demonstrates the complexity in our modern world of all the paperwork surrounding the issuance and completion of a technical directive—an order to a contractor authorizing him to perform work. In this case, a test failure has interrupted

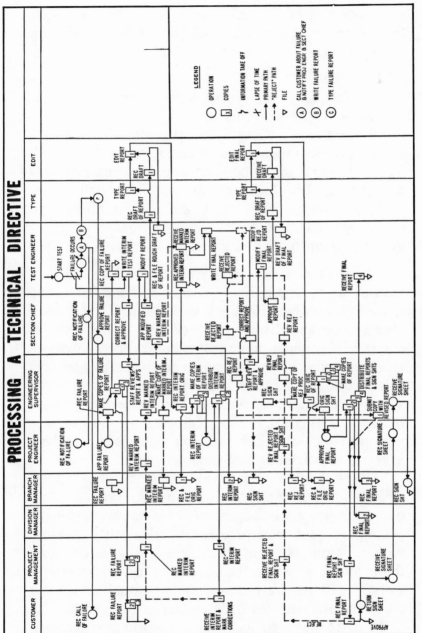

*Fig. 5.8 Paperwork tracing chart*

the processing of the order. Who does what? In brief, the following activities are set into motion, although the actual amount of work involved is far more complicated, as the figure shows:

1. Test Engineer initiates a failure report.
2. Stenographer types it.
3. Section Chief reads and approves.
4. Project Engineer reads and approves.
5. Engineering Supervisor approves and forwards one copy to Test Engineer with recommendations for correction, one copy to Branch Manager for information and filing, and copies to Project Management office for delivery to customer and filing.

A worker effectiveness study can be recorded in various ways, summarizing what was done, how long it took, and what questions were asked. The worker effectiveness and the process itself may be improved as a result of the questioning. Procedures for analysis of repetitive work have been given; for basically nonrepetitive work (for instance, the activities of a secretary), a simple tally sheet will often provide an effective measure of how time is spent. In many cases the operator can use a simple log chart, which will later lend itself to analysis in terms of percentages of time occupied.

*Work Sampling* is a valuable and reliable tool, which allows the analyst to take random samples of an operation being studied. From these he can derive the percentage occurence of each job element, answering such questions as "How much time during the day does the printing press (or the duplicator or the secretary) work?" The rules are many, and it would be wise to study the auxiliary texts mentioned in References 7, 8, and 9 to answer even such simple questions as this one. However, the simple procedure below will often provide significant information.

| DATE | RUN | SET-UP | MAINTAIN | IDLE |
|------|-----|--------|----------|------|
| OCT. 10 | IIII | III | II | I |
| OCT. 11 | ⊬ӀӀ | I | I | III |
| OCT. 12 | III | I | II | ⊬ӀӀ |
| OCT. 13 | II | ⊬ӀӀ | I | II |

*Fig. 5.9    How random samplings are tallied*

First a chart is designed (like that of Fig. 5.9) to show the frequency of the observations. The tally marks indicate individual observations at random times. These are established by dividing the work-day into precise moments (such as 8:00, 8:30, and so forth) and writing them on small cardboard chips. (Lunch hour may be entered as an activity and will give a good check of randomness.) The chips are placed in an opaque container, or their identity otherwise concealed, and ten chips are drawn before the daily study begins. These will list the times at which observations are to be taken. The activity at the *exact* time must be recorded, not an activity just finished or about to begin. About ten samples a day for a minimum of 20 days should be taken in a single study. The observation period must be carefully arranged, moreover, so that the samples can be considered representative. For instance, the activities of clerks during the pre-Christmas season will not be representative of their activities in mid-July. Finally, the tally marks are summarized, making it possible to express the activities recorded from the different observations in percentages of total work time.

The case of a printing press may serve as a typical example. The distribution of the observations during a sampling procedure is indicated in Fig. 5.10. A simple

|  | RUN | SET-UP | MAINTENANCE | IDLE | TOTAL OBSERVATIONS |
|---|---|---|---|---|---|
| NO. OF OBSERVATIONS | 165 | 40 | 30 | 27 | 262 |
| PECENTAGE OF TOTAL | 63% | 15% | 12% | 10% | 100% |

*Fig. 5.10    Result of random samplings*

calculation shows that the press requires 63 percent of the available time and that set-up or make-ready time is 15 percent. Maintenance needs a surprising 12 percent, and idle time represents an objectionable 10 percent. The investigation following the analysis revealed an inadequate ink supply, old and inadequate ink rollers, and poor logistics (the operator had to leave his machine to obtain supplies like paper, ink, hand trucks, and printing plates). In short, the weak points in the operation of the printing press were discovered by the sampling, and a correction of procedures was indicated. The most obvious corrective measures suggested turned out to be these:

1. Reduce maintenance by providing proper ink rollers and ink supply.

2. Reduce downtime by requiring supervision to anticipate logistical problems.

Additional studies—a man movement chart and a work place analysis, for example—might spot additional wasted effort during set-up.

## REFERENCES

1.  American Society of Mechanical Engineers, *Operation and Flow Process Charts*, Standard 101.

2. Bowman and Fetter, *Analysis for Production Management*, rev. ed., 1961, Richard D. Irwin, Inc., Homewood, Ill.
3. Nadler, G., *Motion and Time Study*, McGraw-Hill, New York, 1953.
4. Gilbreth, F.B. and L.M., *Applied Motion and Time Study*, Sturgis and Walton Co., New York, 1917.
5. Nadler, G., *ibid.*
6. Maynard, H.B., Stegemerton, G.J., and Swab, S.L., *Methods Time Measurement*, McGraw-Hill, New York, 1948.
7. Thilgen, G.P., and Procopio, J.F., "Function-Method Approach to Work Sampling," *Journal of Industrial Engineering*, Mar. 1967, p. XV.
8. Barnes, R.M., *Work Sampling*, 2nd ed., Wiley, New York, 1957.
9. Lambrau, Fred H., *Guide to Work Sampling*, Rider, New York, 1962.

# 6

# *Program*
# *Effectiveness Analysis*

Correct timing of the different phases of a project is critical in insuring that bottlenecks and breakdowns do not develop. Rescuing projects that are behind schedule always means excess costs as a result of panic measures, expensive overtime, and other emergency procedures. Coordination may be facilitated by various graphic means. The charts and techniques listed here are used widely in this connection and are particularly interesting and important:

1. The *Gozinto Chart* is a form of the Critical Path Method adapted for relatively small, simple projects where repetition is limited.

2. The *Critical Path Method* (CPM) is a forerunner of PERT. It is not quite as sophisticated as the latter and is therefore preferred by many.

3. The *Program Evaluation Review Technique* (PERT) is one of the best known approaches; it is an excellent base for nonrepetitive programs.

4. The *Line-of-Balance* charting technique—at least 25 years old—is still a superb tool for controlling non-

repetitive programs. It concerns itself with an entire program from raw materials to end products. It relates the actual status of the basic elements of a program to their planned progress, showing exactly how well each is doing. It differs from PERT/CPM in that it is simpler and concerns itself mostly with the production progress of an entire program (usually one consisting of many identical end products).

## 6.1 Gozinto Chart Planning

The Gozinto Chart of Fig. 6.1 demonstrates the steps necessary in planning a Value Analysis Seminar. It is formally called a time phased network and shows: (1) the responsible chairmen, (2) the items on which action is to be taken, and (3) the approximate date on which action must be taken. The "lines of dependency" inform each chairman who will be depending on him to finish his work and on whom he must depend for input. The chart encourages cross-communication and complete understanding of the project.

Figure 6.2 is another example of a Gozinto Chart, of which only a portion is shown. Its purpose is to coordinate the manufacture of a master recorder to be used for recording several incoming sound signals from a noise test range. Usually, the simplest way to plan this complex item would involve planning each step and listing all manufacturing efforts on a "route sheet" with a time schedule added to each sheet so that the various departments and workshops know when to work on the item and how much time is allocated to it. However, such lists become long and cumbersome and the dependence of one item on another is lost in the hundreds of work sheets issued to the numerous departments for the individual items.

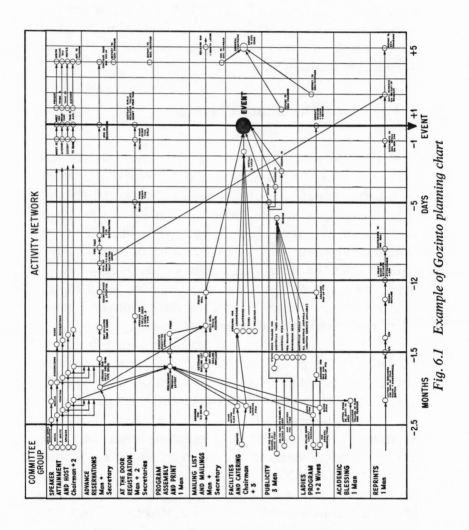

*Fig. 6.1   Example of Gozinto planning chart*

In the Gozinto Chart of Fig. 6.2, each step in the manufacture of the master recorder has been listed. The number of each item is taken from the route sheets, and a "circle" for its assembly is placed below the scheduled date, the latter appearing in chronological order. (The purchase order numbers noted after the parts help to expedite delivery.)

It can be seen immediately that items 030, 031, 036 and 037 would never arrive in time for their scheduled assembly. In fact, 036 and 037 (in short supply) were due to arrive on the job completion date, long after they were required. These facts were learned only after laying out the chart. The arrow pointing to item 057 indicates a seven-day delay from the planned date of receipt, Feb. 20. This delay can be tolerated, since the item will not be used until March 4.

One new concept has been added—the "black line." This is the *critical path*. It represents the route with the greatest number of manhours and therefore the shortest time (as measured in manhours) within which the job can be completed.

The figures at the top of the chart represent the percentage of manhours accumulated to the date shown. They enable the project engineer to state quantitatively the stage of completion the project has reached at any time by adding up all phases that have been completed and deducting those that have not. Completed circles can be marked with a colored pencil for a rapid evaluation of progress.

A simple chart like this, maintained by an aide, has allowed a project engineer to effectively control five times as many shop jobs as he was formerly able to.

A rocket exhaust deflector was successfully completed in six weeks—cutting two weeks off an eight-week schedule—with the help of the Gozinto chart which is shown in Figure 6.3. A page of the route chart

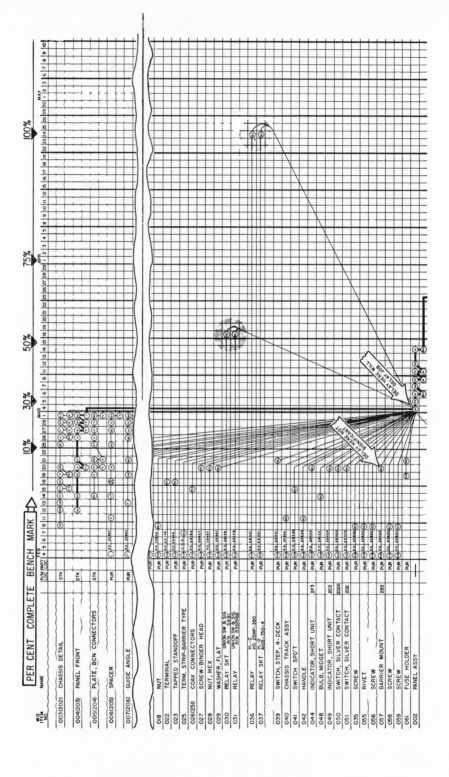

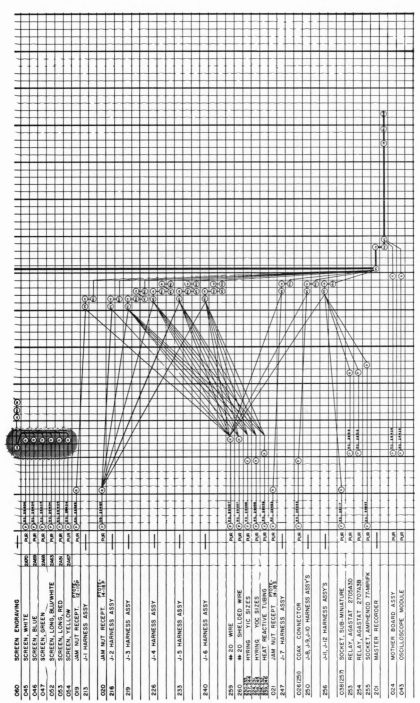

Fig. 6.2   Gozinto chart for a simple project

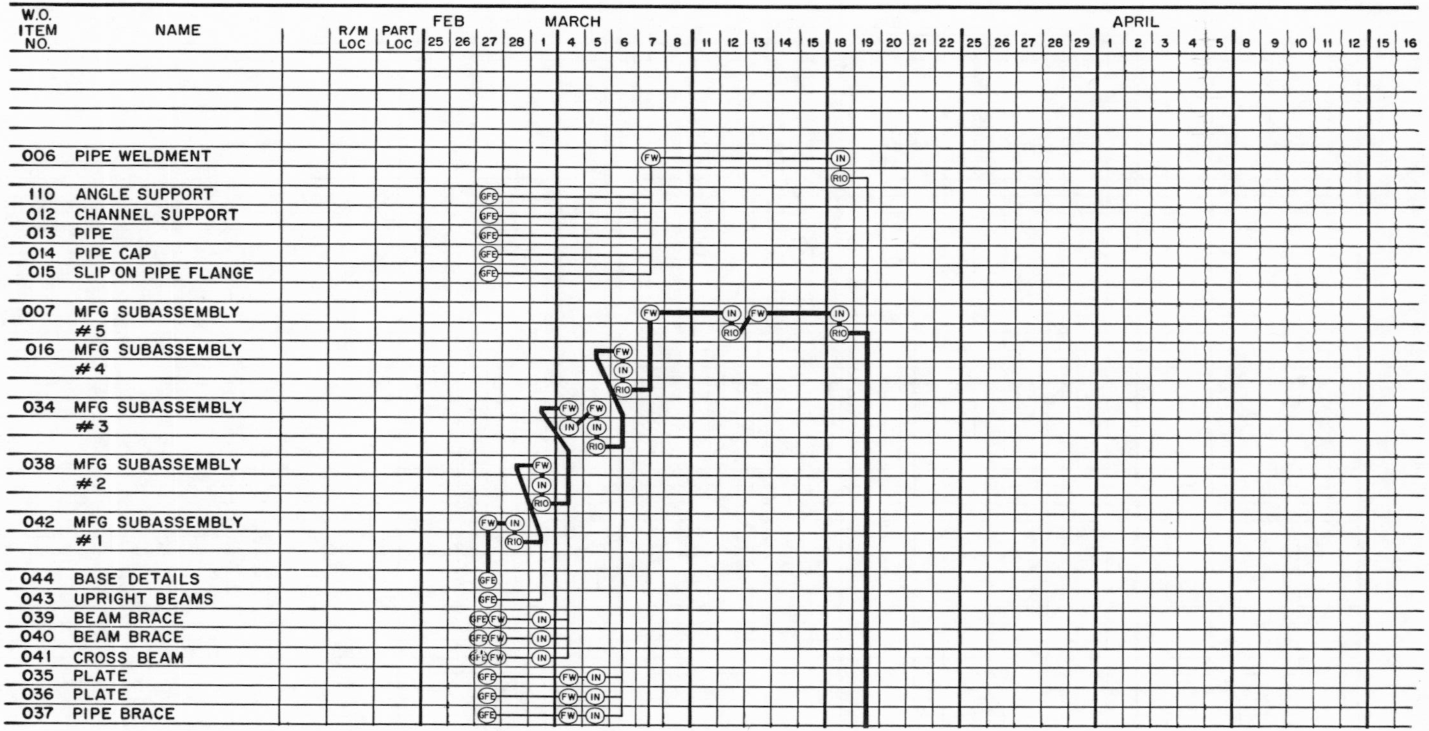

W.O. ITEM NO.
NAME
R/M LOC
PART LOC
FEB
MARCH
APRIL
25 26 27 28 1 4 5 6 7 8 11 12 13 14 15 18 19 20 21 22 25 26 27 28 29 1 2 3 4 5 8 9 10 11 12 15 16
006 PIPE WELDMENT
110 ANGLE SUPPORT
012 CHANNEL SUPPORT
013 PIPE
014 PIPE CAP
015 SLIP ON PIPE FLANGE
007 MFG SUBASSEMBLY #5
016 MFG SUBASSEMBLY #4
034 MFG SUBASSEMBLY #3
038 MFG SUBASSEMBLY #2
042 MFG SUBASSEMBLY #1
044 BASE DETAILS
043 UPRIGHT BEAMS
039 BEAM BRACE
040 BEAM BRACE
041 CROSS BEAM
035 PLATE
036 PLATE
037 PIPE BRACE
FW
IN
RIO
GFE

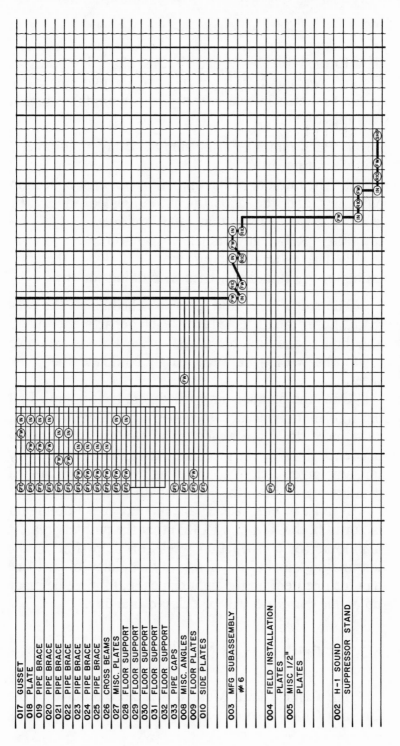

*Fig. 6.3   Critical path*

that formed the backbone of this chart is shown in Fig. 6.4. A route sheet is produced by the planning office as the first step in releasing a job to manufacturing. It lists the following:

1. Each manufacturing step to be taken, in precise manufacturing order.

2. The number of the operation (for cost collection purposes) and the department designated to do the work ("FW," fabrication and welding; "IN," inspection; "RIO," government inspection).

3. The estimated hours required, from the time estimates (or standards) used in bidding for the job.

The route sheet now goes to scheduling. The estimated hours are deducted from each department's available hours and suitable dates—when time is available to do the work—are supplied to complete the route sheet.

Once the planners and schedulers have supplied all the information needed to construct a Gozinto chart, all we need do is to list every item, plot their lines of dependency showing the preceding items that are necessary for any given activity, and finally find the path of greatest manhours—the critical path line.

It must be emphasized again (since it is a dominant factor in both CPM and PERT) that the path of greatest manhours (or construction weeks or shop time) is the dominating single line of activity. A delay of just one day for one item on the tight critical path must be made up or *the entire job* will be delayed one day.

## 6.2    Critical Path Method (CPM)

The Critical Path Method is a close relative of the Gozinto Chart method; both record significant events

BECO FORM  304 A   REV 4/61

| ISSUE DATE | | ISSUED BY | | | | | | |
|---|---|---|---|---|---|---|---|---|

| ORDER NO. 4931- 2182 | | ASSEMBLY NO. TEST B | | DESCRIPTION: MFG (I) H-I SOUND SUPPRESSOR STAND | | | | |
|---|---|---|---|---|---|---|---|---|

| ITEM NO | PART NO AND DRAWING NO | NO REQ | NAME OF PART | MATERIAL DATA | | | | |
|---|---|---|---|---|---|---|---|---|
| | | | | QTY | SHAPE  −  SIZE  −  LENGTH | | TYPE | LOC |

ORDER # 4931-2182   ITEM 002   PAGE 1 OF 1

| OPER NO | DEPT OR MACH | EST HRS | OPERATION | SCH START DATE | SCH COMP DATE | DATE COMP. | OPER. CLOCK NO. |
|---|---|---|---|---|---|---|---|
| 010 | 233 FW | 6 | Layout, Burnout, Fit Up & Weld In | 3/27 | | | |
| | | | Misc ½" Plates As Req'd Per Dwg | | | | |
| | | | Clean Up Welds  (See Sect. F2 On Sht. | | | | |
| | | | 5 For Some Of Plts Req'd ) | | | | |
| 015 | 246 IN | | Process | | | | |
| 025 | R10 | 4 | Ord Insp | 3/27 | 3/28 | | |
| 030 | 233 FW | 6 | Layout & Fab Misc Field Instl Plates | 3/29 | | | |
| | | | (Items 004) (With Excess For | | | | |
| | | | Fit At Assy) Per Dwg. Ref:  See | | | | |
| | | | Plan View Sht 2. Ship Separate. | | | | |
| 040 | 233 FW | 2 | Check Over Assy & Work As Req'd | 3/29 | 3/27 | | |
| 045 | 246 IN | 4 | Final | 3/29 | 4/1 | | |
| 055 | R10 | 4 | Ord Insp | | | | |
| 060 | 233 FW | 8 | Help Load Assy As Req'd | 4/2 | 4/3 | | |
| 500 | 243 IN | 12 | Ship | 4/4 | 4/4 | | |
| | | | | | | | |

*Fig. 6.4    Route sheet used for chart of Fig. 6.3*

as a plan of action against a time scale. To appreciate its value, consider the old-fashioned chart in Fig. 6.5, which was prepared to coordinate the building of a computer center. Its accompanying Gantt chart is shown in Fig. 6.6. Unlike the old-fashioned simple listing of items in

Fig. 6.5, the Gantt chart established general relationships between the various work phases, but the interdependency of these phases and precise procedures continue

| MACHINERY | | | BUILDING | | |
|---|---|---|---|---|---|
| EVENT NO. | DESCRIPTION | DAYS | EVENT NO. | DESCRIPTION | DAYS |
| 0 | PRELIMINARY DISCUSSION | — | 1 | STUDY | 14 |
| 1 | STUDY | 14 | 8 | PRELIMINARY DISCUSSION | 21 |
| 2 | SELECTION | 7 | 9 | FINAL DESIGN | 21 |
| 3 | PLACE ORDER | 3 | 10 | ADVERTISE FOR BIDDERS | 14 |
| 4 | ARRIVAL | 90 | 11 | SELECT CONTRACTOR | 4 |
| 5 | INSTALLATION | 20 | 12 | BUILDING | 76 |
| | | | 13 | FINISH DETAILS | 5 |
| | TOTAL | 134 | | TOTAL | 155 |

| PERSONNEL | | | COMPLETION | | |
|---|---|---|---|---|---|
| EVENT NO. | DESCRIPTION | DAYS | EVENT NO. | DESCRIPTION | DAYS |
| 1 | STUDY | 14 | 6 | INSTALLATION COMPLETE | 7 |
| 14 | DETERMINE REQUIREMENTS | 14 | 7 | PROGRAM | 7 |
| 15 | ADVERTISE | 30 | 18 | DEBUG | 14 |
| 16 | HIRE | 21 | 19 | PRODUCE | 14 |
| 17 | TRAIN | 30 | 20 | "ON STREAM" | — |
| | TOTAL | 109 | | TOTAL | 42 |

*Fig. 6.5   Master plan for construction project*

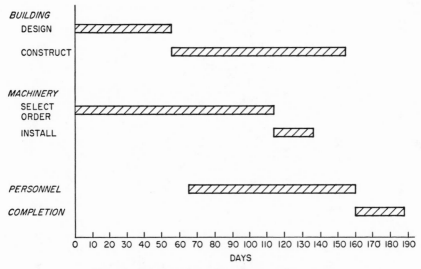

*Fig. 6.6    Gantt chart for activity of Fig. 6.5*

to be vague and individual steps of progress are concealed in huge blocks of effort

The equivalent CPM chart—Fig. 6.7—is laid out in detail against a time scale. Now, if each circle is colored red on completion of the item, one glance at the chart will determine the status of the project. Once used, the effectiveness of CPM will be so evident as to make the old Gantt chart obsolete.

## 6.3    Program Evaluation Review Technique (PERT)

This widely described technique (see References 1 through 4 at the end of the chapter) is an adaption of CPM and Gozinto Charts. It is more sophisticated, though whether it is more effective is debatable.

PERT starts with the same fundamentals as CPM

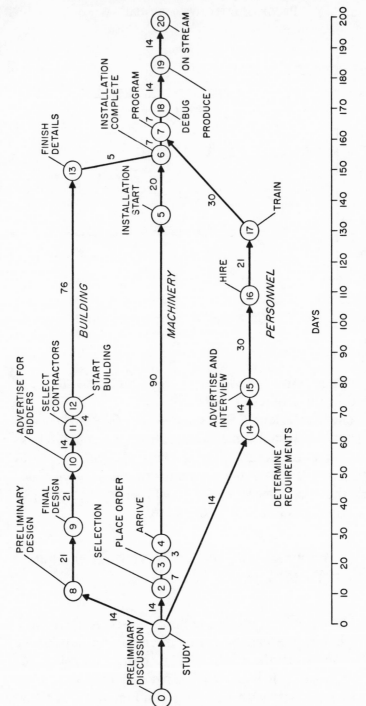

*Fig. 6.7  CPM chart for building activity*

and Gozinto but adds two new features: Three estimates of the completion date are made, and at any time an estimate is available on the probability of meeting the schedule.

The three estimates of the completion date are labeled (1) "most optimistic," (2) "most likely," and (3) "most pessimistic." Between them, they help determine a statistically true critical path, avoiding the errors of fixed or rash judgments. Substituting a range of estimates tends to make the final figure more significant by moving it into the area of statistics.

The expected time to do a job can be found by the following simple formula:

$$\text{Expected Time } (E_t) = (MP + 4ML + MO)/6$$

in which:

MP = "most pessimistic" time
ML = "most likely" time
MO = "most optimistic" time

Expected time is now substituted for the single time used in the CPM chart.

The time scale is often abandoned in favor of showing the elapsed time for each event on the activity line between circles. The critical path can now be calculated by finding the *longest* series of connecting lines or paths. Slack or excess time in companion paths can be found and recorded. For example, consider the following simple example of a series of events, plotted in Fig. 6.8:

1. The preliminary plan
2. Manufacturing plans
3. Electrical plans
4. Manufacturing plans (detail)
5. Procurement
6. Receipt of raw materials

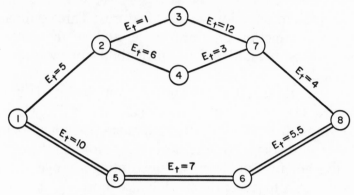

*Fig. 6.8   PERT network chart for simple project*

7. Production plan freeze (begin training)
8. Commencement of production.

The expected time and the slack for each of the paths illustrated is as follows:

| Events | $E_t$ | Slack |
|---|---|---|
| 1–2–3–7–8 | 22 | + 0.5 |
| 1–2–4–7–8 | 18 | + 4.5 |
| 1–5–6–8 | 22.5 | 0.0 |

The most time-consuming, and therefore the critical, path is 1–5–6–8, for which the expected time is 22.5 days. The slack for the other two paths indicates that path 1–2–3–7–8 may be slipped one-half day without effect on the schedule and path 1–2–4–7–8, four and one-half days. One day's slippage on the critical path, however, will mean a slippage of a full day on the estimated schedule.

PERT provides a look at the truly critical items likely to delay a job. For small networks, up to about 500 items, the critical path can be calculated easily by hand. For large and complex programs with thousands of events, it can easily be computerized. Estimating the probability of meeting a schedule is one more contribution of PERT to the management of value.

## 6.4    Line of Balance

The 25-year-old *Line-of-Balance* charting technique for complex nonrepetitive programs is finding new recognition as an excellent tool for keeping top management accurately advised of actual program progress compared to planned progress. An effective cross-sectional review of an entire program, it remains a positive and up-to-the-minute means of determining which areas call for corrective action. Program management can then commit appropriate resources to take care of slippages, improve performance, and reduce costs.

The Line-of-Balance technique takes into account, in graphic form, the essential factors of a program from raw materials to end product. Each factor is carefully numbered to provide for subsequent rapid identification. The sum of these graphic presentations is at any time a measure of the current relationship of production and cost and how they compare to scheduled progress, making it feasible to predict the probability of success.

Figure 6.9 resembles a Gozinto block diagram. It is a chart of the basic building blocks of the equipment being constructed. It diagrams the essential components of the system: many assembled modules, which go into "mother boards," which go into racks, which go into cabinets.

Note that each operation has two numbers. For example, the assembly of the "4-Gate Flip-Flop" starts with number 11 and finishes with number 12. The first

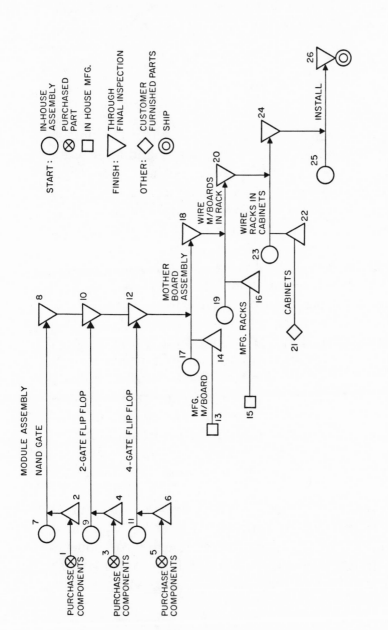

Fig. 6.9  Typical "line-of-balance" network chart

(odd) number is used to identify the date when the operation is to begin (or parts are received) and the second (even) number identifies when it is to be finished (usually at the final inspection).

Figure 6.10 shows the status of all operations as of a specific date (usually posted once a month for long-term projects). The vertical bars, using the same numbers as in Fig. 6.9, show the number of units actually started (odd) and finished (even). The dashed line indicates the number of units planned to be started and finished at the given date. It is this "line of balance" which gives the technique its name. When all the vertical bars touch the planning line, the project is in balance and on schedule.

Figure 6.10 tells us that all module assembly operations (7-8, 9-10, 11-12) are markedly behind schedule (although all purchased parts have been received and inspected), that the manufacture of racks is also behind schedule, and that only the manufacture of boards meets

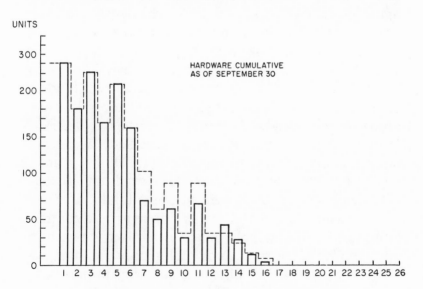

Fig. 6.10   Graph for check on scheduled progress

(and exceeds) its planned progress. The fact that so many even-numbered bars fail to reach the planning line is a positive indication that the program is going to be forced into a time extension unless corrective action is taken.

## 6.5     Line-of-Balance Cost Control

Any worthwhile planning and control tool must consider costs. The Line-of-Balance cost control is as ideally suited for the control of a massive program as a small one. Figure 6.11 provides a simple explanation of how this tool works.

*Step 1:* In the center chart of Fig. 6.11 each distinct cost area to be measured has been laid out on a horizontal bar against a time scale. The cumulative amount of money expected to have been spent by the end of each month is entered on the top of the bar (each figure to be multiplied by $1,000); the percentage it represents of the total is entered below the bar. As the money is spent, the bar is filled in to correspond to the amount, and a "time-now" dashed line is inserted in the scale to allow the viewer to see the current funding status at a glance. For example, at the October 1st date, according to the end of September reports, the Electrical Manufacturing Department had spent $386,000, or 39 percent of its planned total of $1,011,000. The "time-now" line shows that the planned expenditure by that date was $446,000, or 44 percent of the total.

This record is kept for as many areas as need control. Engineering, for example, might be broken down into any number of parts, such as preliminary planning, secondary planning, final planning, checking, project management, testing, and the like.

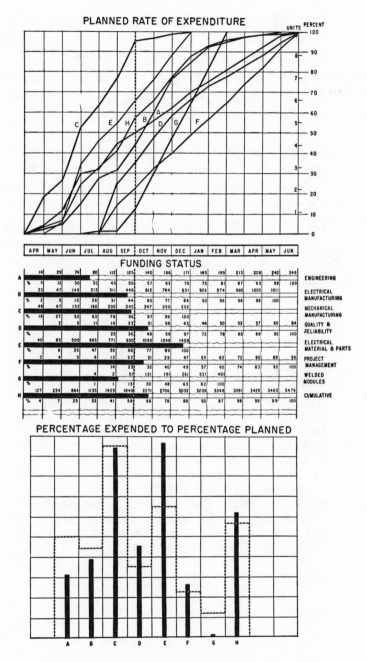

*Fig. 6.11 Line of balance adapted to cost control*

*Step 2:*    The planned rate of expenditure is shown plotted in the top chart of Fig. 6.11. Dollar figures need not show the ordinate, since they are readily available from the lower bars. Percentages are used instead; they will be useful in Step 3. This chart gives a quick visual indication of the rate of fund expenditure.

*Step 3:*    The bottom chart of Fig. 6.11 shows the actual percentage of expenditures as of October 1st against the percentage planned. The percentage planned —the dotted line—is projected from the top chart, where the date line intersects the percentage rate of expenditure.

For more details on the use of the Line-of-Balance technique, see References 5 and 6.

## REFERENCES

1.  Federal Electric Co., *A Programmed Introduction to PERT*, Wiley.
2.  Davis, E.W., "Resource Allocation in Project Network Models," *Survey Journal of industrial Engineering*, April, 1966, p. 177.
3.  *PERT, a Programmed Instruction for Industry*, Basic Systems, Inc. 1963.
4.  Archibald, R.D., and Villoria, R.L., *Network Based Management Systems*, Wiley, 1967.
5.  Finck, Norman E., "Line of Balance Gives the Answer," *Systems and Procedures Journal*, July-Aug. 1965.
6.  *Line of Balance Technology*, Office of Naval Material, Department of the Navy, April, 1962.

# 7

# *Decision-Making Analysis*

In many games the players keep track of the score so that at the end of the play they will have an indication of the relative skill they have displayed. The principles of the social game may also be applied to organized decision making. Alternatives are compared in terms of the "scores" they are able to achieve. The decision-making "game," in other words, can be played for point stakes, with 100 points indicating the optimum alternative.[1]

## 7.1  The Decision Matrix

Matrix analysis, as the decision-making game is called, is easy to play and remarkably versatile, since it can be applied both to simple choices and to complex decisions.

---

[1]  This chapter has been taken from a paper presented at Value Engineering Seminar, University of Alabama, February 22-26, 1965, by D.H. Denholm.

A decision matrix makes it possible to compare alternative decisions. It is a means for the systematic notation, weighting, and evaluation of criteria. The importance of each criterion in relation to total expected performance is defined by giving it a weight or emphasis figure. Each alternative earns a rating based on its weighted performance criteria, expressed in percent; 100 percent, of course, represents the ideal solution.

## 7.2    A Car Buying Example

Figure 7.1 is an example of a simplified decision matrix consisting of an array of vertical columns listing criteria and horizontal columns listing alternatives. Performance figures are weighted in each column and their sum listed in the last column at right. It is obvious that the alternative offering the highest total figure is the winner of the decision.

| | ECONOMY | RELIABILITY | INITIAL COST | TOTAL |
|---|---|---|---|---|
| WEIGHTS | 0.4 | 0.3 | 0.3 | I |
| CAR A | 0.36 (90%) | 0.21 (70%) | 0.21 (70%) | 0.78 |
| CAR B | 0.28 (70%) | 0.27 (90%) | 0.24 (80% | 0.79 |
| CAR C | 0.32 (80%) | 0.24 (80%) | 0.24 (80%) | 0.80 |

*Fig. 7.1    Decision matrix comparing three cars in terms of economy, reliability, and cost*

For the sake of simplicity in this example of the purchase of a car only three criteria are specified : economy, reliability, and price. The matrix shown has been established to coordinate these criteria and make the final choice among the three makes of car—designated A, B, and C—as simple as possible.

Since some criteria may be considered more important than others, it is necessary to give each one a "weight" defining its relative importance. The correct assignment of these weights is the main difficulty in reaching the right decision. Weighting is often biased by a subjective preference for one or the other of the criteria, and this preference may dictate the decision.

In the given case, economy has been given a weight of 0.4, reliability 0.3, and initial cost 0.3, adding up to the required unity. Car A has been given a rating of 90 percent for economy; the weight of this criterion being 0.4, the score of Car A is thus 90 % of 0.4, or 0.36. The other factors are evaluated similarly, the final evaluation showing that car C is the best buy.

To use only three criteria for such a choice is of course far too elementary; normally many more criteria have to be taken into consideration. (Some examples which come readily to mind are the cost of spare parts, appearance, trunk space, and the like.) The probability of making a good choice increases with the number of criteria established and with the number of experienced people working together and making the decision in a team effort. But even in this case weighting the importance of the criteria remains to some degree subjective.

# 7.3    Choosing A Stock Portfolio

A more complex diagram, Fig. 7.2, deals with the selection of stocks on the market today. Though not

| | PAST EARNINGS GROWTH | VALUE RANGE SELL PRICE | CURRENT YIELD | CASH FLOW PER SHARE | DIVIDEND GROWTH | QUALITY RATING | PRICE EARN RATIO | FUTURE OF INDUSTRY | FUTURE OF COMPANY | INSTITUTE HOLDINGS JUNE | RATING |
|---|---|---|---|---|---|---|---|---|---|---|---|
| | FORBES 50%=100 25%=50 0=0 | 0 / 25 / 50 / 75 / 100 | 10%=100 5%=50 0 | 60°SLOPE=100 45°SLOPE=75 30°SLOPE=50 15°SLOPE=25 0°SLOPE=0 | 60°SLOPE=100 30°SLOPE=50 0°SLOPE=0 | INVEST 100 UP MED 75 MED 50 LOW MED 25 SPEC 0 | 0 TIMES=100 100—V 100 TIMES=0 | EXPLOSIVE 100 MED 75 STEADY 50 DECLINE 25 DISASTER 0 | EXPLOSIVE 100 MED 75 STEADY 50 DECLINE 25 DISASTER 0 | 400—100 300—75 200—50 100—25 0—0 | |
| WEIGHT | 15 | 5 | 5 | 10 | 5 | 15 | 5 | 15 | 20 | 5 | 100.00 |
| ALLIED CHEM. | 3.2 (21%) | 3 (60%) | 1.7 (33%) | 0.7 (7%) | 0.15 (3%) | 11.3 (75%) | 4.4 (87%) | 11.3 (75%) | 5 (25%) | 5 (100%) | 45.75 |
| ALUM. CO. | 0 (0%) | 1 (20%) | 0.75 (15%) | 0.6 (6%) | 0.3 (6%) | 11 (75%) | 3.6 (72%) | 7.5 (50%) | 12 (60%) | 3.3 (65%) | 40.05 |
| AMER. CAN | 0 (0%) | 2.5 (50%) | 2.4 (47%) | 0.7 (7%) | 0.15 (3%) | 13.5 (90%) | 4.3 (85%) | 12 (80%) | 16 (80%) | 4.3 (85%) | 55.85 |
| AT & T | 1.7 (11%) | 1.3 (35%) | 1.5 (29%) | 0.7 (7%) | 0.15 (3%) | 15 (100%) | 3.9 (77%) | 11.3 (75%) | 16 (80%) | 5 (100%) | 56.55 |
| AMER. TOB. | 3.3 (22%) | 2.5 (50%) | 2.4 (48%) | 1 (10%) | 0.2 (4%) | 11.3 (75%) | 4.3 (86%) | 6 (40%) | 15 (75%) | 2.5 (50%) | 48.50 |
| ANACONDA | 0 (0%) | 3 (60%) | 2.9 (57%) | 7 (77%) | 0 (0%) | 11.3 (75%) | 4.5 (91%) | 7.5 (50%) | 15 (75%) | 1.3 (25%) | 52.50 |
| BETH. STEEL | 0 (0%) | 3 (60%) | 2.1 (41%) | 3 (30%) | 1.5 (30%) | 7.5 (50%) | 4.3 (85%) | 11.3 (75%) | 15 (75%) | 2.8 (55%) | 50.50 |

Fig. 7.2 Decision matrix based on 10 factors

directly related to VA/VE, this example has been chosen because it demonstrates very well how performance ratings based on past experience may be used to increase the probability of making a correct decision.

The necessary research was provided by an advisory service, which compared 30 stocks showing the greatest strength in the preceding 20 years to determine which of the 30 were the best buys. Of these, only seven are shown in Fig. 7.2. A single matrix indicates growth, performance, and potential. Taken into consideration were ten factors to which carefully selected weights were given. The 10 factors considered important were:

1. Past earnings over the last 10 years.

2. Value range given by the subscribing service (a rather subjective rating).

3. Current yield in terms of dividends paid at the current price.

4. Cash flow per share (based on the amount of money generated by the company in earnings, less depreciation.)

5. Dividend growth

6. Quality rating based on the speculative or conservative character of the company.

7. Price/earnings ratio, or the ratio of the selling price of the stock to the average earnings per share.

8. Industry rating, based on industry growth or decline in comparison with noncompetitive industries.

9. Company outlook.

10. Institutional holdings (the number of professional investing companies that hold shares of stock in each of these companies).

The sum of the weights this time is taken as 100. Of the seven stocks shown, AT&T leads with 56.55 points. The analysis here is more detailed than that of the previous example, but the actual weighting is once again not entirely an objective matter, if only because it can be adjusted to suit specific goals.

## 7.4 Choosing a Supplier

The problem of rating a supplier's ability to meet a company's needs is an important one. Figure 7.3 gives the competitive considerations in buying a bulk chemical from one of three suppliers.[2] As before, weighting re-

| | COST EFFECTIVE-NESS | PURITY | DELIVERY | PRODUCTION FLEXIBILITY | PROXIMITY | FINANCIAL STABILITY | RATING |
|---|---|---|---|---|---|---|---|
| WEIGHT | 20 | 30 | 10 | 20 | 10 | 10 | 100 |
| VENDOR A | 16.0 (80%) | 22.5 (75%) | 8.0 (80%) | 16 (80%) | 7.3 (73%) | 8.5 (85%) | 78.3 |
| VENDOR B | 18.0 (90%) | 26.1 (87%) | 8.0 (80%) | 18.0 (90%) | 8 (80%) | 8 (80%) | 86.1 |
| VENDOR C | 15 (75%) | 21.3 (71%) | 9.0 (90%) | 18.0 (90%) | 8.7 (87%) | 7.5 (75%) | 79.5 |

*Fig. 7.3   Matrix used in selecting a supplier*

flects the relative importance of each criterion; for example, purity is considered three times as important as the ability to deliver. Six competitive factors are assumed to be significant: (1) Cost effectiveness, (2) purity, (3) delivery, (4) production flexibility, (5) proximity, and (6) financial stability of the company.

Percentage ratings were limited to from 70 to 90 percent. A supplier so poorly equipped as to fall below 70 percent wouldn't have been considered, and none

---

[2] Example used by permission of Value Analysis and Purchasing Research, Radio Corporation of America, Camden, New Jersey.

| | LOWER LIMIT OF TOLERANCE ("LEAST WE CAN LIVE WITH") | HIGHEST LIMIT OF TOLERANCE ("MOST WE CAN USE") |
|---|---|---|
| FINANCIAL STABILITY | 90% | 99% |
| PRODUCT COST | $2.50/lb | $2.25/lb |
| PURITY | 99.2% | 99.8% |
| DELIVERY | MONTHLY | EVERY 7 DAYS |
| PEAK QUANTITY REQUIREMENTS | 10,000 lb ON 10 DAYS NOTICE | 50,000 lb ON 5 DAYS NOTICE |
| SUPPLIER PROXIMITY | 2,000 MILES | 100 MILES OR LESS |

*Fig. 7.4    The abstract qualities of Fig. 7.4 reduced to measurable parameters*

were so perfect as to rate 100 percent. Figure 7.4 shows the practical limits of minimum and maximum performance for each factor. The minimum limit may be considered a 70 percent rating; the maximum, a 90 percent rating. Ratings for each factor are established from research tables or utility curves.

Figure 7.5 is a product purity chart that covers the purity variation from 99.2 through 99.8 percent that was established by Fig. 7.4. The former is to be taken as a rating of 70; the latter, a rating of 90. The simplest way to convert the degree of product purity into a rating between these limits is with the straight-line relationship shown. On this scale, Vendor A with a product purity of 99.5 percent would rate 80.0; Vendor B with a product purity of 99.7 percent would rate about 86.7; and Vendor C with a product purity of 99.3 percent would rate about 73.3. But buyer benefits are not always in the linear proportion shown here. A "utility" curve, arbitrarily designed, corrects in this case for marginal utility at the lower levels of purity, penalizing where small variations can cause serious losses. On a corrected curve, Vendor C is rated only 71.5 percent, and this rating has been used.

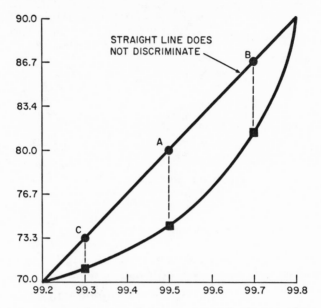

*Fig. 7.5  "Utility curve" for rating vendors (curved line designed to favor best performers)*

## 7.5    The Forced Decision Technique

All rating systems based on distributive weights for value-oriented parameters have one weak point in common: The final choice depends on preliminary personal estimates which are inevitably somewhat subjective and arbitrary and hence not altogether reliable. The distribution of weights also becomes more difficult as more parameters are taken into consideration.

The fundamental virtue of the *forced decision technique* is simplicity, the replacement of one complex decision with a large number of simple decisions: yes—no; more—less; better—worse. The question, "Is A more economical than B?," for example, can be answered more easily than the question, "Should economy be given a value ranging from 70 to 90 percent in a factor weighted at 40 percent of all factors considered?"

An example will show clearly how the forced decision technique is applied in practice, and how weights assigned to a part change if that part is used in different applications.

Consider a part whose basic function is described as "indicate pressure." One application might be as a component of a flight control altimeter; another, a component of a household barometer. To make comparison easy, the evaluation of the part for every application is based on the same parameters: reliability, shock resistance, size, appearance, and cost. (It should be emphasized that parameters which are themselves composed of a number of more limited parameters should be avoided wherever possible. Instead of investigating the merit figures for "reliability," it would be more accurate and realistic to introduce the parameters that constitute reliability: stability, insensitivity to changes in temperature and humidity, corrosion resistance, and so on.)

The first step is to decide the importance of each parameter. A table like that of Fig. 7.6 is set up, in which each of the parameters for a diaphragm used in an altimeter is in turn compared with all the others. The "forced decision" in each case is between "more important" and "less important," represented by the numbers "1" and "0," respectively.

In the first five columns of Fig. 7.6A, sensitivity is compared in turn with reliability, shock resistance, size, appearance, and cost. It is seen to be considered by the team who did the evaluating to be less important than reliability and shock resistance, but nevertheless more important than size, appearance, and cost. Going on to the next factor, reliability, four columns are devoted to comparing it with shock resistance, size, appearance, and cost (it has already been compared with sensitivity).

| | 1 | 2 | 3 | 4 | 5 | 6 | 7 | 8 | 9 | 10 | 11 | 12 | 13 | 14 | 15 | n | n/15 |
|---|---|---|---|---|---|---|---|---|---|---|---|---|---|---|---|---|---|
| SENSITIVITY | O | O | I | I | I | | | | | | | | | | | 3 | 0.200 |
| RELIABILITY | I | | | | | I | I | I | I | | | | | | | 5 | 0.333 |
| SHOCK RESISTANCE | | I | | | | O | | | | I | I | I | | | | 4 | 0.266 |
| SIZE | | | O | | | | O | | | O | | | I | O | | 1 | 0.066 |
| APPEARANCE | | | | O | | | | O | | | O | | O | | O | 0 | 0.000 |
| COST | | | | | O | | | | O | | | O | | I | I | 2 | 0.133 |
| | | | | | | | | | | | | | | | | 15 | 1.000 |

(A) DIAPHRAGM USED IN ALTIMETER

| | 1 | 2 | 3 | 4 | 5 | 6 | 7 | 8 | 9 | 10 | 11 | 12 | 13 | 14 | 15 | n | n/15 |
|---|---|---|---|---|---|---|---|---|---|---|---|---|---|---|---|---|---|
| SENSITIVITY | I | I | O | O | O | | | | | | | | | | | 2 | 0.133 |
| RELIABILITY | O | | | | | I | O | O | O | | | | | | | 1 | 0.066 |
| SHOCK RESISTANCE | | O | | | | O | | | | O | O | O | | | | 0 | 0.000 |
| SIZE | | | I | | | | I | | | I | | | O | O | | 3 | 0.200 |
| APPEARANCE | | | | I | | | | I | | | I | | I | | O | 4 | 0.266 |
| COST | | | | | I | | | | I | | | I | | I | I | 5 | 0.333 |
| | | | | | | | | | | | | | | | | 15 | 1.000 |

(B) DIAPHRAGM USED IN DOMESTIC BAROMETER

Fig. 7.6   Forced decisions on a barometric device

Each of the other factors is evaluated similarly, and the number of positive decisions for each ($n$) is totalled at the right to get its weight of importance. This weight, or "emphasis coefficient," is arrived at by dividing $n$ by the greatest possible number of positive decisions. In this case the emphasis coeffiicient is seen to be $n/15$.

The same procedure is followed in Fig. 7.6B. In this case the diaphragm is to be used in a less sophisticated domestic appliance, a barometer, in which cost and appearance are dominant concerns. Comparing the $n$-columns of the two tables, one notes that the altimeter evaluation emphasizes reliability, shock resistance, and sensitivity (weights 5, 4, and 3, respectively) whereas for the domestic barometer the heaviest weights (also 5, 4, and 3) have been given to cost, appearance, and size.

Now that the weights for each parameter have been allocated, a similar matrix (Fig. 7.7) is set up to compare the various types of barometer or altimeter in terms of these qualities. The piston, bellows, metal diaphragm, silicon-rubber diaphragm, and mercury barometer types are compared with each other, first with respect to sensitivity, then to the other parameters. The decisions are again made in terms of "1" or "0," and are independent of the application (altimeter or barometer) since each of the devices is being checked against a given quality. For instance, the bellows is more sensitive than the piston, but the piston's cost is lower. The number of positive decisions is added up and the emphasis coefficient determined as in the calculations for weighting in Fig. 7.6.

To determine the final merit figures, the emphasis coefficients for each type of the device must now be multiplied by the weights for the particular parameters. To this end, the matrices of Figs. 7.8 and 7.9 are set up. Each of the parameters has a column, and each type of device has two rows. The top row shows the emphasis

| Item | 1 | 2 | 3 | 4 | 5 | 6 | 7 | 8 | 9 | 10 | n | n/10 | Characteristic |
|---|---|---|---|---|---|---|---|---|---|---|---|---|---|
| PISTON | 0 | 0 | 0 | 0 | | | | | | | 0 | 0.0 | SENSITIVITY → |
| BELLOWS | 1 | | | | 1 | 0 | 1 | | | | 3 | 0.3 | |
| METAL DIAPHRAGM | | 1 | | | 0 | | | 0 | 0 | | 1 | 0.1 | |
| SILICON-RUBBER DIAPHRAGM | | | 1 | | | 1 | | 1 | | 1 | 4 | 0.4 | |
| MERCURY BAROMETER | | | | 1 | | | 0 | | 1 | 0 | 2 | 0.2 | |
| PISTON | 0 | 0 | 0 | 0 | | | | | | | 0 | 0.0 | RELIABILITY → |
| BELLOWS | 1 | | | | 1 | 1 | 1 | | | | 4 | 0.4 | |
| METAL DIAPHRAGM | | 1 | | | 0 | | | 1 | 1 | | 3 | 0.3 | |
| SILICON-RUBBER DIAPHRAGM | | | 1 | | | 0 | | 0 | | 0 | 1 | 0.1 | |
| MERCURY BAROMETER | | | | 1 | | | 0 | | 0 | 1 | 2 | 0.2 | |
| PISTON | 0 | 0 | 1 | 1 | | | | | | | 2 | 0.2 | SHOCK RESISTANCE → |
| BELLOWS | 1 | | | | 1 | 1 | 1 | | | | 4 | 0.4 | |
| METAL DIAPHRAGM | | 1 | | | 0 | | | 1 | 1 | | 3 | 0.3 | |
| SILICON-RUBBER DIAPHRAGM | | | 0 | | | 0 | | 0 | | 1 | 1 | 0.1 | |
| MERCURY BAROMETER | | | | 0 | | | 0 | | 0 | 0 | 0 | 0.0 | |
| PISTON | 0 | 0 | 0 | 1 | | | | | | | 1 | 0.1 | SIZE → |
| BELLOWS | 1 | | | | 0 | 0 | 1 | | | | 2 | 0.2 | |
| METAL DIAPHRAGM | | 1 | | | 1 | | | 0 | 1 | | 3 | 0.3 | |
| SILICON-RUBBER DIAPHRAGM | | | 1 | | | 1 | | 1 | | 1 | 4 | 0.4 | |
| MERCURY BAROMETER | | | | 0 | | | 0 | | 0 | 0 | 0 | 0.0 | |
| PISTON | 0 | 0 | 0 | 1 | | | | | | | 1 | 0.1 | APPEARANCE → |
| BELLOWS | 1 | | | | 1 | 1 | 1 | | | | 4 | 0.4 | |
| METAL DIAPHRAGM | | 1 | | | 0 | | | 1 | 1 | | 3 | 0.3 | |
| SILICON-RUBBER DIAPHRAGM | | | 1 | | | 0 | | 0 | | 1 | 2 | 0.2 | |
| MERCURY BAROMETER | | | | 0 | | | 0 | | 0 | 0 | 0 | 0.0 | |
| PISTON | 1 | 0 | 0 | 1 | | | | | | | 2 | 0.2 | COST → |
| BELLOWS | 0 | | | | 0 | 0 | 1 | | | | 1 | 0.1 | |
| METAL DIAPHRAGM | | 1 | | | 1 | | | 1 | 1 | | 4 | 0.4 | |
| SILICON-RUBBER DIAPHRAGM | | | 1 | | | 1 | | 0 | | 1 | 3 | 0.3 | |
| MERCURY BAROMETER | | | | 0 | | | 0 | | 0 | 0 | 0 | 0.0 | |

Fig. 7.7  Characteristics compared in two's, with better-worse choices represented by 0 and 1

| PARAMETER | SENSITIVITY | RELIABILITY | SHOCK RESISTANCE | SIZE | APPEARANCE | COST | n DECISIONS |
|---|---|---|---|---|---|---|---|
| POSITIVE DECISIONS | 3 | 5 | 4 | 1 | 0 | 2 | 15 |
| EMPHASIS COEFFICIENT | 0.200 | 0.333 | 0.266 | 0.066 | 0.000 | 0.133 | 1.000 |
| PISTON | 0 / 0 | 0 / 0 | 0.2 / 0.0532 | 0.1 / 0.0066 | 0.1 / 0.00 | 0.2 / 0.0266 | 0.0864 (5) |
| BELLOWS | 0.3 / 0.06 | 0.4 / 0.1332 | 0.4 / 0.1064 | 0.2 / 0.0132 | 0.4 / 0.0 | 0.1 / 0.0133 | 0.3261 (1) |
| METAL DIAPHRAGM | 0.1 / 0.02 | 0.3 / 0.1 | 0.3 / 0.0798 | 0.3 / 0.02 | 0.3 / 0.00 | 0.4 / 0.0532 | 0.2730 (2) |
| SILICON RUBBER DIAPHRAGM | 0.4 / 0.08 | 0.1 / 0.0333 | 0.1 / 0.0266 | 0.4 / 0.0264 | 0.2 / 0.00 | 0.3 / 0.0399 | 0.2062 (3) |
| MERCURY BAROMETER | 0.2 / 0.04 | 0.2 / 0.0666 | 0 / 0 | 0 / 0 | 0 / 0 | 0 / 0 | 0.1066 (4) |

*Fig. 7.8 Emphasis coefficients and merit numbers established for altimeter*

| PARAMETER | SENSITIVITY | RELIABILITY | SHOCK RESISTANCE | SIZE | APPEARANCE | COST | n DECISIONS |
|---|---|---|---|---|---|---|---|
| POSITIVE DECISIONS | 2 | 1 | 0 | 3 | 4 | 5 | 15 |
| EMPHASIS COEFFICIENT | 0.133 | 0.066 | 0 | 0.200 | 0.266 | 0.333 | 1.000 |
| PISTON | 0 / 0 | 0 / 0 | 0.2 / 0 | 0.1 / 0.02 | 0.1 / 0.0266 | 0.2 / 0.0666 | 0.1132 (4) |
| BELLOWS | 0.30 / 0.0399 | 0.40 / 0.0264 | 0.40 / 0 | 0.20 / 0.04 | 0.40 / 0.1064 | 0.10 / 0.0333 | 0.2460 (3) |
| METAL DIAPHRAGM | 0.10 / 0.0133 | 0.30 / 0.020 | 0.30 / 0 | 0.30 / 0.06 | 0.30 / 0.0798 | 0.4 / 0.1332 | 0.3063 (1) |
| SILICON RUBBER DIAPHRAGM | 0.40 / 0.0532 | 0.10 / 0.0066 | 0.10 / 0. | 0.40 / 0.08 | 0.20 / 0.0532 | 0.3 / 0.1 | 0.2930 (2) |
| MERCURY BAROMETER | 0.20 / 0.0266 | 0.20 / 0.0132 | 0 / 0 | 0 / 0 | 0 / 0 | 0 / 0 | 0.0398 (5) |

*Fig. 7.9  Domestic barometer figures*

coefficient for the device, and the bottom row the product obtained by multiplying that number by the weighting coefficient at the head of the column. Thus in the case of sensitivity, the mercury barometer for the altimeter (Fig. 7.8) has an emphasis coefficient of 0.2. This figure multiplied by the weighting coefficient of 0.200 at the top of the column gives a product of 0.04.

Finally, the merit figures are obtained by adding the products for all parameters (bottom line for each device). These merit figures are listed in the last column, with their relative standings in parentheses still further right. A fast check for the correctness of the calculation is to add all the merit figures, which should add up to 1.

The results show that the bellows or metal diaphragm is the appropriate choice for the altimeter application (merit figures of 0.3261 and 0.2730 respectively), whereas the metal diaphragm or the silicon-rubber diaphragm (merit figures of 0.3063 and 0.2930) is the best solution for the domestic application.

## REFERENCES

1.  McKinsey, J.C.C., *Introduction to the Theory of Games*, McGraw-Hill, 1952.
2.  Fasal, John H., *"The Forced Decision for Value,"* *Product Engineering*, April 12, 1965.

# 8

# *Predictability and Probability*

The value engineer is always in a better position when he can supply management with reliable goals. Round numbers like 5 or 10 percent are too often assumed to be definite objectives; management is actually saying, "We need to reduce costs—what can you do?" If the value engineer has a technique for predicting the results of his engineering effort, he can propose a series of tradeoffs in performance and cost; for example, reducing costs by 10 percent while maintaining performance or improving performance 5 percent while holding costs. Management will then be better equipped to cope with the problem of cost reduction because it will then be aware of the possible alternatives.

It is equally necessary for the value engineer to be knowledgeable about the allocation of resources to all items in a system so that he can develop a sense of proportion in regard to value versus cost. A single item may be responsible for 50 percent of the cost of a system but be functionally responsible for 90 percent of the performance. It would not make sense to reduce the cost of this item, and yet some value engineering techniques are directed at items of high costs with no appreciation of value versus performance tradeoffs.

The value engineer has to contend with two allocation problems: the first is the allocation of the resources for building an item or system; the second is the allocation of value engineering efforts so as to ensure the most effective use of those resources. The analysis of the first problem (allocation of resources) is of primary importance, since its solution must be used to solve the second one.

Before, during, and after the time the value engineer expends his efforts to improve the performance/cost relationship of an item, the question may arise whether these efforts could not be spent more profitably. Although the value engineer may have saved $1000, could he have saved $2000 with the same effort applied to another aspect of the problem? While there can never be a completely satisfactory answer to that question, with the application of certain techniques the value engineer can be reasonably confident that he is applying his efforts to the problems that will give the highest rewards. In other words, he can predict the maximum results of his efforts. This prediction may be based on intuition, on a simple arithmetic formula, or on the application of sophisticated principles of probability. All three approaches, which have met with both success and failure, will be discussed in this chapter.

## 8.1    General Considerations

Consider the problem of a value engineer analyzing a hypothetical system which has been value engineered before to some unknown degree. If the value engineer is experienced and has worked on similar systems, he will know intuitively where to start his analysis to get the largest payoffs. In large, complex systems, intuitively knowing where to start requires a great deal of experience. The value engineer must be able to analyze all parts of a system for performance and cost, establish a perspective, and then select the areas that should receive the greatest attention for the best initial return on his efforts. The same method can obviously be used to decide which of several systems would benefit the most from VE efforts.

When the engineer is inexperienced or lacks a "feel" for changes in performance—cost relationships, or when the system is highly complex or new, a more formal technique is required. It is then preferable to make an assessment of one or more systems, estimate the work and results, and select an area of concentration based on the highest payoff. This payoff can be expressed as:

$$\text{Payoff} = \frac{\text{Results of value engineering efforts}}{\text{Cost of value engineering efforts}}$$

The results of the value engineering efforts are assessed in terms of cost savings or performance improvements. With the usual limitations on time, and skills, it behooves the value engineer to be able to predict the payoffs on each item in the system.

It is relatively easy to determine the payoff after a project has been completed, but before starting a project, predicting the payoff is difficult. It requires a system analysis and (preferably) a correct mathematical model of the performance cost relationship for each item in the system. By having a mathematical model of his system, the value engineer can see where to apply his efforts to change the performance cost ratio most favorably, and he can determine the impact of that improvement—if any —on other parts of the system. This model will help him reallocate the resources assigned to the program to attain the highest cost effectiveness, thus providing new performance and cost goals for each item.

## 8.2      Allocation of Resources

To appreciate the proper allocation of resources, it is necessary to understand the worth of the product. The worth consists of the resources that go into the product less the company's indirect expenses and profit. The ideal of highest performance, lightest weight, smallest size, and lowest cost cannot be realized. There are always compromises or tradeoffs dictated by the customer's requirements or other factors. These tradeoffs are of primary importance in the allocation of resources.

The theories of value and worth have already been explained. For the purpose of this chapter, it can be stated that there is a value of use and a value of exchange. In the mind of the purchaser, the value of an item is based either on its performance (or potential performance) or merely on its effect. For example, a diamond ring may not perform a function, but it can instill pride of ownership in an individual or be sentimentally rewarding. This pride of ownership or sentiment is worth something, and if the price is right the individual will buy the diamond ring. Therefore, in the following discussions, value and cost have to be considered as being independent, and value will be taken as the measure of performance per unit cost.

It is difficult to allocate money to research projects of unknown potential, since the performance (hence value) of an as yet unknown quantity is obviously unknown. Many project managers allocate their resources on an essentially intuitive basis. However, more astute technical managers will assign a value—regardless of how arbitrary it may be—based on the knowledge of whether the task is feasible. For example, a project manager who needs to develop a special adhesive may settle for a small project to determine whether a certain class of materials is applicable, rather than on the large project that would be required to actually find the adhesive. The knowledge gained from the small project is of tangible value; the problem is to determine exactly what this value is.

During the last fifty years, the technique of scientific management has caused managers to depend less on "feel" and more on predictable results. The allocation of resources has progressed from the work of Gantt on the scheduling of work to PERT/CPM. The key feature of these efforts consists of examining both the estimated effort and the results concurrently and resolving the problem of the proper utilization of manpower by careful scheduling. The procedure (often used without knowing it is a scientific management technique) consists of breaking a job down into its smallest component activities (within reason), and then estimating the amount of effort required by each. They are then ordered in the necessary sequence; those efforts that are independent are paralleled; and the most critical activities are noted.

The control of the job has evolved into a philosophy of management by exception: those activities that are critical, that are giving the most trouble, or that are late in performance are managed most closely. The emphasis is on control, with the allocation of resources being made by experience and trial-and-error. High-cost operations or parts are identified and efforts made to reduce them. Thus has involved the classical value engineering technique of breaking a project or system down into its smallest functional elements, and, by relating function to cost for each, improving the performance/cost relationship of the whole.

Many top managements lack reliable criteria for selecting value engineering projects. Studies have revealed that most value engineering projects are selected with little or no concern for maximizing the return on the investment of value engineering resources; reducing the high cost of individual items (rather than the total costs as related to the functional performance of the system) is the usual objective. Although it has been shown that value engineering is most effective during research and development, as little as a quarter of such efforts have been expended before design release.

A simple (if not optimal) technique has worked well and produced improvements. Sometimes called the ABC method, it consists of ranking all of a company's products in terms of the total dollar cost of a year's production. (The rank of a product is also based to some degree on quality and criticality, along with other considerations.) Experience has shown that 10 to 15 percent of the products account for 60 to 80 percent of the costs. This is called category A. Category C consists of 60 to 80 percent of the items representing 10 to 15 percent of the costs. Category B falls in the middle. The ABC method of selecting value analysis projects consists in starting at the top of category A and working down into category B (depending on the VA resources available). The secret of the technique is in knowing when to quit.

As each item is developed to improve its cost effectiveness, the gains are recorded, and when value engineering costs approach gains too closely it is time to quit. In practice, of course, the procedure is not quite that simple. In general, the VA team works on several projects simultaneously, and cost effectiveness must be gauged in terms of VA costs for the total effort. A cost/ effectiveness ratio for the value-engineering group has to be set, for instance, 8 dollars worth of gain per dollar expended on VA time. When the ratio falls much below this figure, the VA effort becomes unfeasible.

The ABC technique can be modified by the use of weighting factors for each item; this often results in moving a B item into the A category. The Navy has developed the Numerical Value Rating System (NVRS), which uses nine parameters: number of parts, quantity of operations in manufacture, function, complexity, weight, material, man-hours required, reliability, and maintainability. These parameters were constructed so that the NVRS number would vary directly with cost. Other approaches, as, for instance, the "forced decision" technique, have already been discussed.

All these techniques are an aid in selecting a value engineering project. However, since it is not easy to assess the value of an item with respect to the system, the value engineer requires a system analysis using a mathematical model in order to be able to predict the payoff.

## 8.3     Programming Techniques

There are two basic categories of systems, and nearly all operating entities fall into one of them; some fall into both. Those in the first category are designated *series configurations*. These systems operate when their subsystem functions are performed sequentially, are interdependent, and are not substitutable. (For example, the carburetor of an automobile cannot be substituted for the transmission.) Those in the second category are designated *parallel configurations*, and in these the various subsystems operate independently. (For example, in a machine shop each machine is a subsystem in which many functions and operations are parallel.) This distinction between configurations is made because resources in each case are allocated differently; it is important to understand the distinction to avoid a misapplication of techniques.

In a machine shop with several mills, lathes, and the like (each having different functions and operating costs), the problem is to determine which machine should be used to manufacture which items for minimum cost. The shop is a parallel system in which any of several machines could conceivably be allocated to a product. This system can be examined (or math modeled) with linear programming, an operations research technique. Each machine, however, requires the functioning of several parts in sequence (such as motor, bed, tool, table, and so on) and is therefore a series subsystem. The allocation of resources to these elements of the machine cannot be done optimally with linear techniques.

Two types of mathematical models have been developed for allocating resources on a systems basis. The first type consists of *programming techniques* (linear programming) that are applicable to systems in a parallel configuration. These techniques were first successfully applied on a large scale to the massive logistics problems caused by World War II and require the development of a matrix in which the resources are enumerated in rows and the demands or performances in columns. The result is a series of simultaneous equations that are solved by matrix algebra.

Linear programming assumes that all interrelationships are linear. Nonlinear programming assumes that the cost/objective function is nonlinear and requires the matrix solution to be repeated as the conditions change. Dynamic programming introduces a time variable and requires successive solutions with respect to time. For small problems the programming techniques require a disproportionately great effort for the results obtained. For large, complex problems, the programming techniques require a computer and involve a considerable amount of set-up time, but in many such cases programming offers the *only* solution to the problem. When programming is to be used, the value engineer should be prepared to expend considerable effort.

The second type of model for allocation of resources was developed more recently for application to series configurations. It is referred to as the simplified method and is based on the concept that performance (utility) is related to cost in an exponential manner. This relationship has been known for some time.

A series system is analyzed so that the performance of each item in it is correlated to the over-all system performance; each has a cost effectiveness of its own with respect to the system. Since the performance/cost relationship is exponential, the cost effectiveness will change with the budget assigned to the item. The goal is to arrive at a budget for the system in which the cost effectiveness of each item is identical.

The advantages of the simplified method are that it is easy to use and can be applied to any series type configuration; it can predict the performance of an item, for a given budget, with respect to the overall system performance; and it establishes target performances and budgets. The disadvantages are that it is not applicable to parallel configuration systems, and the data required for the analysis is, in general, difficult to obtain.

## 8.4    Linear Programming

After World War II the United States Air Force became aware of the potential usefulness of various scientific programming techniques in coordinating the efforts of the whole nation efficiently in the event of a global war. The generalized form of this approach to organization problems resulted in the "linear programming models" which industry has adopted to its great advantage since about 1950.

Linear programming is one of the newer mathematical disciplines used to aid management in decision making. As business firms and their operations continue to increase in size and complexity, management continues to be confronted with new problems and new uncertainties. The proper use of linear programming techniques can indicate the decisions most likely to be economically beneficial to a company.

Optimization problems are those that seek to maximize or minimize a function of a number of variables which have certain defined constraints. In theoretical economics, in government, the military, business, and industry, many new and vital optimization problems associated with linear programming do not yield to classical methods of optimization. It was necessary to formulate a new method to solve these problems efficiently. Mr. G. B. Dantzig is credited with the development of one, the Simplex method, the fundamentals and working principles of which will be explored further in this section.

A linear programming problem requires, in general, the solution of $m$ linear equations in $n$ unknowns:

$$a_{11}x_1 + a_{12}x_2 + \ldots a_{1n}x_n = A_1$$
$$a_{21}x_1 + a_{22}x_2 + \ldots a_{2n}x_n = A_2$$
$$a_{m1}x_1 + a_{m2}x_2 + \ldots a_{mn}x_n = A_m \qquad (8.1)$$

The unknowns are designated $x_1, x_2, x_3 \ldots x_n$; all other quantities are known constants. Indeterminacy is present for $m < n$ and also for $m = n$ if the system of equations is linear dependent. In addition, it is required that

$$x_j \geqq 0 \text{ for } j = 1, 2 \dots n \qquad (8.2)$$

and that a distinct set of values, $x_1, x_2 \dots x_n$, that satisfies Eq. 8.1 maximizes or minimizes the solution of a third equation,

$$Z = c_1x_1 + c_2x_2 \dots c_nx_n \qquad (8.3)$$

which is called the *objective function*. The values of $c_1$, $c_2$, and so forth, are given constants.

Equation 8.1 represents the system's constraints; Eq. 8.2 establishes that all variables remain positive; Eq. 8.3 represents the objective function, the maximizing or minimizing of which is the purpose of the process. This formulation of the linear programming problem is in the standard form, where all the constraints are written as equations, all variables are required to be non-negative, and the objective function is to be maximized or minimized. The word "linear" means that all the constraint equations and the objective function are linear or can be approximated by first-order equations.

The fundamental characteristics of a linear programming problem are (1) an *objective* to be realized (such as minimum cost, maximum profit, or minimum elapsed time, for the system under study) and (2) a large number of variables to be considered simultaneously (these may be products, machine hours, man hours, storage space, required dollars, or the like). Variables of different types may occur in a given problem; for example, manufactured products are *system outputs*, whereas machine-hours or man-hours are *system resources*. There are many possible interactions between the variables required to define the system.

A typical problem—to determine the proper product to be manufactured during a given production period, with limited available machines, so as to secure maximum overall profit—will demonstrate some possible methods of solution. This very simple problem presents little difficulty—the solution is almost obvious—and it was selected for precisely that reason, so that the reader may perceive and understand the manipulations involved in linear programming. The graphical method will illustrate the problem, showing how the objective function of maximum profit is realized; the algebraic method will demonstrate the concept and use of inequalities; the Simplex method will present a typical example of linear programming manipulations adaptable to larger numbers of products and machines.

A manufacturer wishes to produce two products (P1 and P2) during unused time in two machine centers (MC1 and MC2). Up to 80 hours of unused time is available in MC1 and up to 64 hours in MC2; both products must pass through both machine centers. Each unit of product P1 requires 8 hours in MC1 and 4 hours in MC2; the profit per unit is $10. Each unit of product P2 requires 4 hours in MC1 and 8 hours in MC2, and the profit per unit is $6. The total profit (from P1 and P2) is

$$Z = c_1x_1 + c_2x_2 = (\$10)\,(x_1) + (\$6)\,(x_2) \quad (8.4)$$

in which c = contribution to profit made by each unit and x = number of units. Subscripts 1 and 2 refer to products P1 and P2.

The constraints in the problem are the number of machine-hours required to produce a unit of P1 or a unit of P2, and the number of hours available in each machine center. The constraint equations are

$$a_1x_1 + a_2x_2 \leqq A_1 \quad (8.5)$$

and

$$a_2x_1 + a_1x_2 \leqq A_2 \quad (8.6)$$

in which a = the number of hours required to produce each unit, x = the number of units, and A = the total number of hours available. For MC1 we thus have the equation,

$$8x_1 + 4x_2 \leqq 80 \qquad (8.7)$$

For MC2, the equation is

$$4x_1 + 8x_2 \leqq 64 \qquad (8.8)$$

The above equations (or inequalities) must be satisfied, and the objective function (the overall profit), $Z = 10x_1 + 6x_2$, must be maximized. Maximization requires that values of $x_1$ and $x_2$ (which are controlled by the above constraints) be found which will give a maximum value of Z.

## 8.5     The Graphical Solution

This solution is applicable in cases with two variables (three-variable problems can sometimes be solved, though usually only with great difficulty). The following demonstration will help to visualize the concept of simultaneous solutions.

The set of points that satisfies Eqs. 8.5 and 8.6 can be given a graphical representation by allowing the equals sign to hold in each case rather than the "less than" sign. The constraint equation for MC1 becomes

$$8x_1 + 4x_2 = 80$$

or

$$x_2 = (80 - 8x_1)/4 = 20 - 2x_1 \qquad (8.9)$$

A plot of this equation gives the line CD in Fig. 8.1. All points on this line will satisfy the equation, and all points below it will satisfy the inequality.

The constraint equation for MC2 becomes

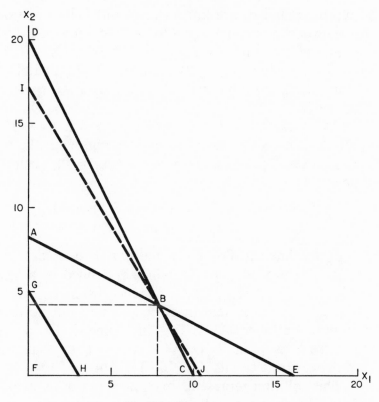

*Fig. 8.1   Graphic solution*

$$4x_2 + 8x_2 = 64$$

or

$$x_2 = (64 - 4x_1)/8 = 8 - (x_1/2) \qquad (8.10)$$

A plot of this equation gives the line AE. This constraint limits solution points to those on or below the line AE. The region of feasible solutions for this problem thus lies within or on the boundary of the area shown by the lines FA-AB-BC-CF.

The objective function for this problem is Eq. 8.4. Since $Z = c_1x_1 + c_2x_2$,

$$x_2 = -(c_1/c_2)x_1 + (1/c_2)Z$$

in which $c_1$ and $c_2$ are known constants. The slope of the objective function is equal to $-(c_1/c_2)$ regardless of the value of Z. In this case, the value of the slope is $-10/6$, or $-5/3$. Any line having this slope determines pairs of variables $x_1$ and $x_2$ that will satisfy Eq. 8.4 for a specific value of Z.

As an example, the line GH of Fig. 8.1, its $-5/3$ slope determined by its intersection points with the $x_1$ and $x_2$ axes ($x_1 = 0$, $x_2 = 5$; $x_1 = 3$, $x_2 = 0$), defines a profit of

$$Z = c_1x_1 + c_2x_2 = 10(0) + 6(5) = 30$$
$$= 10(3) + 6(0) = 30$$

While this line satisfies the equations, it does not represent the maximum profit, which is defined by a line of slope $-5/3$ at the greatest distance from the origin of the coordinate system that is capable of intersecting a point on periphery ABC. This line is line IJ. The point it intersects is point B ($x_1 = 8$, $x_2 = 4$), so that the optimized profit equals $Z = 10(8) + 6(4) = \$104$.

This solution represents maximum use of all available machine time as well as maximum profit but these do not necessarily coincide. If the unit profit, on P1 had been sufficiently larger than that on P2, say $20 for P1 and $6 for P2, maximum profit would coincide with maximum production of P1 and no production at all of P2. MC1 would be used for the full 80 hours, but MC2 would be used for only 40 hours. The value of $x_1$ (10) that will bring about this situation is determined by setting $x_2$ equal to 0 in Eq. 8.9.

## 8.6    The Algebraic Method

In general, it is much easier to work with equations than with inequalities. To convert from inequalities to equations a new variable known as a *slack variable* is added to each inequality to form an equation. In this case the slack variable added to each equation represents the unused time in each machine center. The problem is defined as follows: Maximize the objective function $Z = 10x_1 + 6x_2$ while satisfying the inequalities,

$$8x_1 + 4x_2 \leqq 80$$
$$4x_1 + 8x_2 \leqq 64$$

Let $x_3$ represent the slack variable associated with MC1. It represents the total time available in MC1 (80 hours) less any hours used in that center to process products P1 and P2. Likewise, let $x_4$ represent the total time available in MC2 (64 hours) less any hours used in that center to process products P1 and P2. The inequalities above may now be written as equations:

$$8x_1 + 4x_2 + x_3 = 80 \qquad (8.11)$$
$$4x_1 + 8x_2 + x_4 = 64 \qquad (8.12)$$

Mathematically the values of $x_3$ and $x_4$ are

$$x_3 = 80 - 8x_1 - 4x_2 \qquad (8.13)$$
$$x_4 = 64 - 4x_1 - 8x_2 \qquad (8.14)$$

If no profit or loss is charged against idle machine time, the slack variables for the objective function (profit) can be set at zero:

$$Z = c_1x_1 + c_2x_2 + c_3x_3 + c_4x_4$$
$$Z = c_1x_1 + c_2x_2 + (0)x_3 + (0)x_4$$
$$Z = c_1x_1 + c_2x_2 \qquad (8.15)$$

Note that if profit or loss *is* charged to idle machine time, the slack variables $c_3$ and $c_4$ are to be taken in terms of dollars *per hour* rather than *per unit* as for $c_1$ and $c_2$.

The next step is to find how profits can be maximized. Since product P1 makes the greatest profit contribution per unit, let us assume first that all the time in both machine centers is used for its manufacture. A quick calculation shows that 10 units (80/8) can be made in MC1 and that 16 units (64/4) could be made in MC2 were it not that MC1 limits the total output.

Substituting $x_1 = 10$ and $x_2 = 0$ in Eqs. 8.13 and 8.14, we obtain

$$x_3 = 80 - 8(10) - 4(0) = 0 \qquad (8.16)$$
$$x_4 = 64 - 4(10) - 8(0) = 24 \qquad (8.17)$$

We see that there is no unused time in MC1 and 24 hours of unused time in MC2.

The objective function Z now depends on the variables:

$$x_1 = 10 \text{ (units of product P1)}$$
$$x_2 = 0 \text{ (units of product P2)}$$
$$x_3 = 0 \text{ (unused hours in MC1)}$$
$$x_4 = 24 \text{ (unused hours in MC2)}$$

Consequently,

$$Z = 10x_1 + 6x_2 + 0(x_3) + 0(x_4)$$
$$= 10(10) + 6(0) + 0(0) + 0(24) = \$100 \qquad (8.18)$$

There is no assurance, however, that this value of Z is the optimum one, only that it reflects maximum use of the controlling machine center MC1 for the more profitable product, P1. If a certain number of unused hours in MC2 were given to product P2 and the number of hours devoted to product P1 in MC1 were correspondingly altered, the value of Z might possibly increase. We thus need to find values for $x_1$ and $x_4$ in terms of the other two variables, $x_2$ and $x_3$. The value of Z can then be determined strictly on the basis of these two variables.

It is possible to express $x_1$ in terms of $x_2$ and $x_3$ by using Eq. 8.13:

$$x_3 = 80 - 8x_1 - 4x_2$$

$$8x_1 = 80 - 4x_2 - x_3$$
$$x_1 = 10 - (1/2)x_2 - (1/8)x_3 \qquad (8.19)$$

Similarly, $x_4$ can be expressed as follows (from Eq. 8.14):

$$
\begin{aligned}
x_4 &= 64 - 4x_1 - 8x_2 \\
&= 64 - 4[10 - (1/2)x_2 - (1/8)x_3] - 8x_2 \\
&= 64 - 40 + 2x_2 + (1/2)x_3 - 8x_2 \\
&= 24 - 6x_2 + (1/2)x_3 \qquad (8.20)
\end{aligned}
$$

These values may now be substituted in the objective function to determine whether or not the profit function can be improved:

$$Z = 10[10 - (1/2)x_2 - (1/8)x_3] + 6x_2 + (0)x_3 \\ + (0)[24 - 6x_2 + (1/2)x_3]$$

and the new form of the profit function becomes:

$$Z = 100 + x_2 - (5/4)x_3 \qquad (8.21)$$

The negative coefficient of $x_3$ indicates that any machine hours not used in MC1 will decrease profit and that, ideally, $x_3$ should be zero. The positive coefficient of $x_2$ indicates that it is possible to increase profits by making some of that product. However, the coefficient (unity) indicates that the increase in profit will be only $1 per unit of $x_2$ instead of the original $6.

The next step is to determine how many units of P2 must be made to maximize the profit. Because all the time in MC1 is already being utilized, it will be necessary to make one less unit of P1 for every two units of P2. There are 24 hours of unused time in MC2. Each unit of P2, moreover, requires 8 hours in MC2. But for each unit of P2 made (and a corresponding unit of P1 *not* made), two hours, or one-half the time required to process a unit of P1, becomes available in MC2. Thus, the time required to process a unit of P2 in MC2 becomes 8 − 2, or 6 hours per unit. Since the unused time in MC2 is 24 hours, 4 units of P2 can be made. (To make one more unit of P2, requiring 8 hours of MC2, the number of units of P1 would have to be reduced by two, a loss of $20 against a gain of $6.) If four units of P2 are made in MC1, requiring 16 hours, 64 hours are left for P1, or time for eight units. Since $x_3$ is thus zero and $x_2$ is established at a value of 4, it is clear from Eq. 8.21 that the maximum profit is $104.

The solution is not always so apparent as it is in this simple problem. In such cases, Eqs. 8.19 and 8.20 may be used to find the quantity to be added. The rule is to divide the constant in each equation (10 in Eq. 8.19 and 24 in Eq. 8.20) by the coefficient of variable $x_2$ (1/2 and 6) and to choose the smaller positive quotient as the quantity to be added. As 10 divided by 1/2 is 20, and 24 divided by 6 is 4, obviously 4 is the correct number to add. Substituting 4 for $x_2$, the equations become:

$$x_1 = 10 - 1/2(4) - 1/8(0) = 8$$
$$x_4 = 24 - 6(4) + 1/2(0) = 0$$

With 4 units of P2, 8 units of P1 will be made, and unused time in both MC1 and MC2 is reduced to zero. The maximum profit becomes

$$Z = 10(8) + 6(4) + 0 + 0$$
$$= 80 + 24$$
$$= \$104$$

The algebraic method of linear programming can be used with a number of variables, whereas the graphic method is—for all practical purposes—limited to two.

## 8.7     The Simplex Method

When a large number of equations are involved in a problem, graphical and algebraic methods become impractical and burdensome. For the solution of such problems, the Simplex Method preserves many of the assets of those two techniques, yet eliminates many of the disadvantages. A matrix is formulated and one solution after another is sought, each with a greater objective function (profit). All solutions with objective functions lower than those of previous solutions are eliminated automatically. The process is repeated until no solution with a greater objective function can be found. For a full understanding of this method, the reader is referred to References 11, 12, and 13 at the end of this chapter.

## 8.8    Nonlinear Programming

Many methods have been suggested for the solution of a nonlinear programming problem. (Philip Wolfe in *Recent Advances in Mathematical Programming* discusses a number of computational procedures used to solve this type of programming.) The characteristics of problems that demand the use of nonlinear programming techniques are:

1. The objective cost function represents a nonlinear relationship.

2. The constraint equations are nonlinear with respect to the variables in the problem.

Linear programming techniques have been developed to a very sophisticated level. The development of procedures for solving the nonlinear problem, on the other hand, has only recently reached the point where computational procedures have been possible. The field is very much in its infancy at this time and it appears that further research must be forthcoming. Readers interested in studying this subject further are referred to References 14, 15, and 16 at the end of this chapter.

## 8.9    Summation

It was stated previously that two allocation problems face the value engineer. The first is the allocation of resources to a program; the second, the allocation of the value engineer's own resources. The effectiveness with which the value engineer solves the second problem will decide how well he solves the first. Therefore, it is advisable for him to develop techniques to make his value engineering efforts worthwhile. To do so, it is necessary to understand the various techniques available for allocating resources:

1. Intuition
2. Simplified method based on experimental relation of cost to performance
3. Scientific scheduling (Gantt)
4. PERT/CPM
5. Programming (linear, etc.)

Of these, the intuitive method offers the value engineer no hope. Value engineering improvements have been made by a simplified numerical rating system, but possible optimization cannot be proved or disproved. The Gantt scheduling and PERT/CPM methods offer some information for rating the importance of various elements in the value engineering analysis.

The programming techniques developed in operations research have permitted optimization of complex parallel systems. Although sometimes awkward and difficult to use, they are a powerful method of analyzing a system. It is again strongly suggested that persons desiring to use these techniques plan to spend a considerable time in studying them. In cases where a large number of equations are used, a computer is a must.

# REFERENCES

1. Milles, L.D., *Techniques in Value Analysis and Engineering*, McGraw-Hill, 1961, p. 25.
2. Kao, John H.K., "Statistical Models in Mechanical Reliability," 11th National Symposium on Reliability and Quality Control, 1965, p. 240.
3. *Value Analysis—Value Engineering*, Falcon, W.D., ed., American Management Association, 1964.
4. Petrochell, R.L., and Chester., J.M., *Total Systems Cost Analysis* (RM-3069-PR), Rand Corp., 1963.
5. Gantt, Henry L., *Organizing for Work*, Harcourt Brace, 1919.
6. Clark, Charles E., *The Optimum Allocation of Resources During the Activities of a Network*, ASTIA Report AD 289.417.
7. Dale, E., *Management Theory and Practice*, McGraw-Hill, 1965.
8. *Criteria for Selection of Value Engineering Projects*, VE Research Study Tech. Rpt. No. 3, ATTN Industrial Management Dept., Rock Island Arsenal, Ill.
9. *Numerical Value Ruling System Handbook*, RRBN2-1 NWSA Code RREN-21, U.S. Navy, Washington, D.C.
10. Llewellyn, Robert W., *Linear Programming*, Holt, Rinehart and Winston, 1964.
11. Bellman, Richard E., *Mathematical Optimization Techniques*, University of California Press, 1963.
12. Fabrysky, W.J., and Torgerson, P.E., *Operations Economy*, Prentice-Hall, 1966.
13. Dantzig, C.B., Orden, A., and Wolfe, P., *Generalized Simplex Method for Minimizing a Linear Form for Linear Programs*, Rand Corp., 1954.
14. Rosen, J.B., "The Gradient Projection Method for Non-Linear Programming," *Journal of the Society for Industrial and Applied Mathematics*, Sept. 1966.
15. Kuhn, H.W., and Tucker, A.W., *Proceedings of the Second Berkeley Symposium on Mathematical Statistics and Probability*, University of California Press, 1951.
16. Leontief, W.W., *The Structure of the American Economy, 1919, 1929*, New York: Oxford University Press, 1951.

# 9

# *Statistical Evaluation of Design*

Product value is sometimes defined as the ratio of worth to cost:

$$V = W/C$$

Traditionally, value engineering has concentrated on the denominator—cost—by defining function, challenging requirements, and seeking lower cost alternatives. But cost, in the conventional sense, is not the only factor that the value engineer must consider. As stated in earlier chapters, delays in delivery often make the savings brought about by a lower-cost design insignificant. A production line stopped for one hour can represent the loss of hundreds of dollars; nonavailability of a twenty-cent item can hold up a major system, tying up funds, manpower, and space.

Through value engineering efforts true value is further enhanced by improving quality and reliability —the numerator in the equation—especially by making the product work the first time it is put into production. Management simply cannot allow the production line to be used as an extension of the engineering laboratory. Even worse than this setback, though, is continued engineering once the product is in the field. Warranty costs, retrofits, and quick fixes at this stage not only bite deeply into profits, but they damage corporate reputation and that most precious of commodities, customer satisfaction.

Unfortunately, conventional quality control and reliability programs do not provide the answer. Quality control starts much too late in product evolution—generally not until the start of production. Reliability departments often spend so much time *measuring* reliability that they cannot concentrate on improving it. The result is only too often endless theoretical negotiation about what is and what is not a failure.

The statistical techniques described in this chapter enable value engineer and design engineer to evaluate a design in its early stages, so that not only actual failures but latent defects can be discovered in the breadboard and brass-board phases of product development. Once design is stabilized, the value engineering objective should be to assure that no potential failures are introduced through purchased materials, faulty workmanship, or processes contributing to inconsistencies in duplicating the design. Such an evaluation of design, materials, workmanship, and processes is not only good value engineering; it is good quality control and true reliability; it is the quintessence of a Zero Defects program; above all it is good management.

## 9.1    Statistical Tools

Every defect has a root cause which may not be evident but which is nonetheless effective. An undesired effect may be separated from its root cause by long chains of cause and effect. Traditionally, the root cause of an undesired effect has been discovered intuitively (engineering judgment or hunches). This intuitive approach often works, but it is undependable. For one thing, when the engineer thinks he has, by intuition, seen the source of a problem, he may very well not bother to turn the assumed cause "on" and "off" to see if the effect appears and disappears. But usually the intuitive approach doesn't work for a simpler reason than the failure to test what may be a premature conclusion: the engineer cannot put his finger on the root cause of a defect because he does not know where to begin looking among all the possible causes. There are just too many of them possible.

It is on such occasions of bafflement that a systematic statistical search designed to make the problem "tell on itself" can most profitably replace the hit-or-miss conventional approach. This systematic search is schematically represented by Fig. 9.1, a road map of statistical tools in design and production; the purpose of each statistical tool and its area of application are given in Table 9.1.

The advantages of this battery of statistical tools in problem-solving at any stage of design or manufacture are as follows:

1. Time savings:  a week's statistical experimentation could conceivably solve a problem that a company had been unsuccessfully attacking for years, or which it had given up fighting, accepting it as a "quirk of the process."

2. Cost savings:  statistical diagnosis can give quick solutions, often with a ratio of savings to implementation costs far higher than the standard value engineering figure of 10:1.

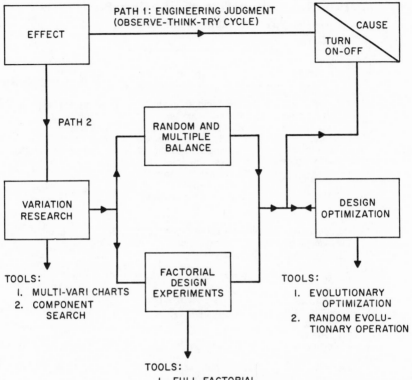

*Fig. 9.1   Road map of troubleshooting techniques*

### Table 9.1   Statistical Techniques, Their Purpose and Area of Application

| Statistical technique | Purpose | Area of application |
|---|---|---|
| *Variation research* Multi-vari charts Component search | Narrow the possible causes of largest variation when traditional engineering methods fail to determine these causes | Prototype or production stage |

**Table 9.1 Statistical Techniques, Their Purpose and Area of Application (Cont'd)**

| Statistical technique | Purpose | Area of application |
|---|---|---|
| *Factorial design* Full factorial Fraction factorial Latin square Hyper square | Pin-point important and interacting variables and measure the contribution of each; open up tolerances and controls on rest of these variables with little effect on observed output (used where number of input variables does not exceed 4) | Brass-board or prototype or production stage |
| *Random and multiple balance* | Same as above (used where number of input variables exceeds 4) | Bread-board or brass-board stage |
| *Design optimization* Evolutionary optimization Random evolutionary operation | Optimize design by moving empirically to a "best" combination of interacting variables | Bread-board or brass-board stage |
| *Overstress* | "Smoke out" weaknesses in design | Prototype stage |
| *Multiple environmental analysis* | Determine consistency of production workmanship and purchased materials | Pilot production and production stages |
| *Statistical tolerancing, scatter plots, sequential plots* | Study variation in production and introduce cost-cutting measures | Prototype and production stages |
| *"High time" experiments* | Evaluate designs that have accumulated considerable field time | Field stage |

3. Minimum disruption: Only a small number of units are required for the statistical experiments, resulting in a minimum of disruption of normal product flow in engineering or in production.

## 9.2    The Observe-Think-Try Cycle

The traditional approach to experimentation is the observe–think–try cycle. First, all aspects of the problem in which an effect occurs are observed. Then, the many causes that might be responsible for a given effect are contemplated, and the most likely cause or factor is tried. If it proves not to be the root cause, the next most likely factor is tried, and so on.

There are serious drawbacks to this approach, even aside from the length of time needed to vary one factor at a time. The results indicate only primary effects (those in which there is a reasonable linear relationship between cause and effect). Interaction effects (those in which several factors in conjunction lead to an effect which no one of them causes by itself) are not detected. (A well-known example of an interaction effect is the human reaction to a tranquilizer pill and alcohol. Neither taken by itself need be dangerous, but the two taken together may be fatal.) Such experimentation may also fail to disclose clues to residual effects—those produced by factors other than the ones considered (common examples are experimental error and errors that creep in with time).

When the traditional common sense approach fails, statistical tools like variation research can prove their value. General principles shared by statistical approaches and in which they differ from the traditional one, may be stated as follows:

1.  Every effect has one or more contributing causes.
2.  These causes do not contribute equally to the effect.
3.  Only a very few causes are responsible for the major portion of the effect. This fact reflects the Pareto-

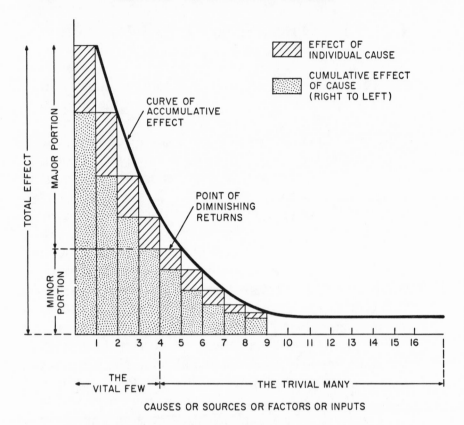

*Fig. 9.2    Pareto-Lorenz maldistribution curve showing contribution of causes to total effect*

Lorenz maldistribution principle widely used in industry and depicted in Fig. 9.2. It indicates that only a few of a large number of contributing causes (or factors or inputs) account for over 50 percent of the total effects. A general rule of thumb is that 20 percent or less of them account for 80 percent or more of the total effects. This principle is widely used in value engineering and bidding, to concentrate on a few high-cost parts; in quality control and reliability, to concentrate on the few most important causes of defects; and in management, to concentrate on the most important products or schedules (critical path).

4. These few causes are not constant in their effects.

5. Techniques to analyze the variation of the total effect give a clear signal, separating the few vital causes from the trivial many.

## 9.3    Variation Research

Variation research follows the pattern of the game of "Twenty Questions," in which the questioner starts with all-inclusive categories such as "animal, mineral, or vegetable" and then narrows the field of inquiry with more limited questions to isolate the specific answer. The two statistical tools that ask the first of those "Twenty Questions" are named the "multi-vari chart" and the "component search."

*Multi-vari chart*     This statistical tool programs tests in terms of three broad categories: (1) variation from unit to unit, (2) variation from place to place within units, and (3) variation from time to time. Thus measurements are made in three different locations within each of three units manufactured in quick succession; this procedure is repeated in three separate time periods, giving 27 measurements of the characteristic under investigation. The results can be plotted as shown in Fig. 9.3.

Let us say that the characteristic being investigated is color saturation. The degree of saturation for each of three locations in three separate units is represented in Fig. 9.3 by the position of the "x." The nearer the "x" to the top margin, the closer the degree of saturation to that which had been specified, and vice versa. It is quickly seen that the greatest variation is from place to place. In places 1 and 3 the variation is fairly consistent, but

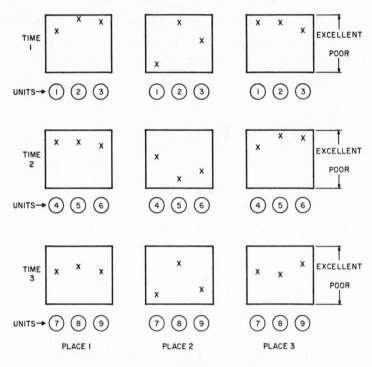

*Fig. 9.3   Multi-Vari chart*

in place 2 it fluctuates wildly. The time variation (apart from place 2) is much less marked, although it does suggest that the third time the test was performed the degree of saturation was somewhat inferior to the first two times.

In an experiment to determine the cause of poor solder connections on printed circuit boards, three broad causes were suspected:

1. *Location within the board*, based upon factors such as hole size, hole density, circuit configuration, board warpage, and board positioning

2. *Differences between successive boards*, based upon factors such as board contamination, different

suppliers, circuit plating, and short-term wave-solder machine-variations

3. *Long-time differences*, based upon factors such as nonuniform temperatures, nonuniform solder composition, acidity in flux, and wave-solder machine maintenance.

Three board locations, A, B, and C were chosen. On each of three separate days, three consecutive units which had been sent over the wave-soldering machine were then analyzed around regions A, B, and C for defective solder connections. The number of defects per 100 connections was plotted as shown in Fig. 9.4. The results indicated that while there was some variation in the percentage of poor solder connections according to

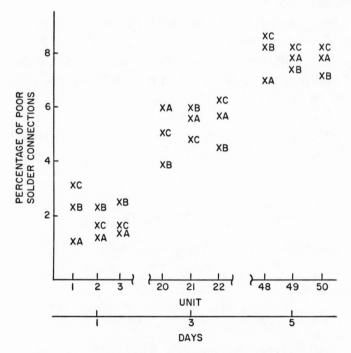

*Fig. 9.4   Application of the Multi-Vari chart for locating causes of poor soldered connections*

board location (A, B, and C), and also slight variation between successive units, it was the variation over time (approximately 2 to 7 percent) that was the most significant. Thus, time-to-time variations were responsible for the poor connections rather than board differences or different locations within a board. The causes of these variations could be further investigated with other statistical tools.

*Component search*     When the multi-vari analysis indicates a large unit-to-unit variation, and where there are definitely "good" units and "bad" units, each containing interchangeable components or subassemblies, the technique called "component search" is often used to isolate the culprit subassembly and to discover its malfunction.

The approach is a series of tabulated experiments—called "2 × 2 factorial designs"—based on the "good" and "bad" units already known. As an example, if one of the components that is suspect is designated as $C_g$ and $C_b$ in the good and bad units, respectively, and the rest of the components as $R_g$ and $R_b$ in the same good and bad units, a 2 × 2 factorial design would depict the original condition as shown in Table 9.2.

**Table 9.2     2 × 2 Factorial Designs**

| Rest of unit | Suspect component | |
|---|---|---|
| | $C_g$ | $C_b$ |
| $R_g$ | Good unit as received | *Component to be verified* |
| $R_b$ | *Rest of unit to be verified* | Bad unit as received |

As an example, in an electronic circuit using a filter, several good units were produced and also several rejects. A typical good unit had an output voltage of 15 volts, whereas a typical bad unit measured 10 volts. It was desired, through component search experiments, to determine if the rejects were caused by the bad filter or by the rest of the circuitry other than the filter. The filter was designated F and the rest of the circuitry as R. The subscript "g" was used with the good unit and the subscript "b" with the bad unit. To determine if the filter was indeed the cause of the rejection, the suspect filter from the bad unit was connected to the good unit and the filter from the good unit connected to the bad unit.

*Result 1*      If the exchange gave voltage readings as indicated in Table 9.3A, it could be assumed that the trouble did not come from the suspect filter but from the remainder of the unit.

To confirm this, the two units should be restored to their original condition, that is, the suspect filter should be returned to the bad unit and vice versa, and the measurements repeated. This is known in design experimentation as "replication." The object of such replication (or repetition) is to measure the effect of experimental error or the residual effect of other causes not considered in the experiment. Thus, if Table 9.3B resulted, it could be safely assumed that there was no change from the original readings and that no experimental error had crept in. In other words, a component other than the filter has to be hoped for as the culprit.

*Result 2*      If the exchange gave readings as shown in Table 9.3C, it could be assumed that the suspect filter was indeed the cause of the trouble. The replication effected by return of the exchanged components to their original condition also confirms that there was no experimental error.

### Table 9.3    Component Search

|       | $F_g$ | $F_b$ |
|-------|-------|-------|
| $R_g$ | 15    | —     |
| $R_b$ | —     | 10    |

Original condition

|       | $F_g$ | $F_b$ |
|-------|-------|-------|
| $R_g$ | 15    | 15    |
| $R_b$ | 10    | 10    |

(A) Result 1

|       | $F_g$    | $F_b$ |
|-------|----------|-------|
| $R_g$ | 15 15    | 15    |
| $R_h$ | 10       | 10    |

(B) Result 1 with Replication

|       | $F_g$    | $F_b$   |
|-------|----------|---------|
| $R_g$ | 15 15    | 10      |
| $R_b$ | 15       | 10 10   |

(C) Result 2 with Replication

|       | $F_g$    | $F_b$   |
|-------|----------|---------|
| $R_g$ | 15 15    | 13      |
| $R_b$ | 12       | 10 10   |

(D) Result 3 with Replication

|       | $F_g$    | $F_b$   |
|-------|----------|---------|
| $R_g$ | 15 15    | 15      |
| $R_b$ | 15       | 10 10   |

(E) Result 4 with Replication

|       | $F_g$    | $F_b$   |
|-------|----------|---------|
| $R_g$ | 15 13    | 13      |
| $R_b$ | 12       | 10 15   |

(F) Result 5 with Replication

*Result 3* If the exchange produced the readings of Table 9.3D and replication confirmed them, the conclusion would be that the filter was only partially responsible for the trouble and that other components should be exchanged to determine the other factors.

*Result 4* If the exchange produced the readings shown in Table 9.3E, with the trouble vanishing following the exchange, it could be assumed that the original problem was caused by the way in which the filter and the rest of the circuitry interacted in the bad unit. An impedance mismatch could cause such an interaction.

*Result 5* If the exchange produced the readings shown in Table 9.3F, with nonrepeatable results in restoring the equipment to its original condition, the conclusion is strong that there is experimental error, caused by unstable equipment, poor measurement techniques, improper assembly, or the like. These extraneous factors, then, should be investigated rather than the units themselves.

The important characteristic of component search is to recognize the number of different patterns (minimum of five in a 2 × 2 design, and more if several interchanges are combined into one test) that can exist. Interchanges have been used for years, but only with the thought of a simple main effect as a rule and generally with the objective of just identifying the bad unit. (If the bad unit improved with a component from the good unit, it was possible that the bad unit would be repaired and shipped without considering whether the root cause had been isolated.)

## 9.4 Factorial Design Experiments

Further solutions may be achieved by factorial design experiments. Such experiments are either observed visually to detect trends or computed mathematically, using analysis-of-variance calculations. The reader is referred to the voluminous statistical literature for details on the use of factorial designs. Only the simplest, the "Latin square" and "random and multiple balance," which are widely used by engineers, will be discussed here.

*Latin square*     This is a design experiment where the number of major suspected factors is always three and the number of levels (classes) of each factor is always equal. The square design is called "balanced" because each level of each factor is used the same number of times with each level of every other factor.

Table 9.4 shows the simplest Latin square; each factor (A, B, C) has two levels (1, 2). Four experiments are to be run: $A_1$ with $B_1$ and $C_1$ for the first experiment, $A_1$ with $B_2$ and $C_2$ for the second, and so on. Since only four of the eight possible experiments are called for, this is analagous to 50 percent sampling and therefore subject to error. However, because major causes tend to appear readily in experiments of this sort, the validity of the result is far greater than with a 50 percent sample.

**Table 9.4    The Simplest Latin Square**

|       | $A_1$ | $A_2$ |
| ----- | ----- | ----- |
| $B_1$ | $C_1$ | $C_2$ |
| $B_2$ | $C_2$ | $C_1$ |

To reduce experimental error caused by setup, measurement, the passage of time, and other extraneous reasons, each of the four experiments should be run twice (replicated).

The technique is best illustrated with a case study example. Field complaints indicated many incidents of a broken cathode-ray tube in a certain complex electronic device when it was removed from the shipping carton for installation. The customer's engineers were convinced that it was a problem of tolerances. The manufacturer of the electronic device suspected an alignment problem. The manufacturer's glass engineer insisted that the defective tubes had scratches in the neck that caused breakage at that point. All were convinced that the damage had not been caused in transit, and this fact was confirmed by a comparison of the percentage of defective devices at the manufacturer's shipping point and on arrival at the installation point.

A Latin square experiment was designed to pinpoint the true problem: glass was the first factor (variable), with two levels (with or without neck scratches); alignment was the second factor, with two levels (perfect alignment or misalignment); tolerance was the third factor, with two levels (minimum or maximum differential between the O.D. of the tube and the I.D. of the yoke). Since each combination of factors was to be run twice (to reduce experimental error), there were eight tests involving eight tubes and eight electronic devices into which the tubes were inserted. In each test, the number of cycles of shock until breakage occurred was recorded. The results are shown in Table 9.5. They may be summarized as follows:

$T_1$ = minimum play or tolerance differential
$T_2$ = maximum play or tolerance differential
Left-hand column of readings   = 30
Right-hand column of readings = 36
Top row of readings             = 32
Bottom row of readings          = 34

**Table 9.5    Latin Square Test of an Electronic Device**

(Numbers within each box indicate the number of cycles before breakage each time the test was run.)

| Alignment | Glass | |
|---|---|---|
| | *With scratches* | *Without scratches* |
| *Misaligned* | $T_2 < \begin{array}{c} 3 \\ 4 \end{array}$ | $T_1 < \begin{array}{c} 13 \\ 12 \end{array}$  } 32 |
| *Perfect* | $T_1 < \begin{array}{c} 12 \\ 11 \end{array}$ | $T_2 < \begin{array}{c} 6 \\ 5 \end{array}$  } 34 |
| | 48    30 | 36    18 |

$T_1$ diagonal of readings        = 48
$T_2$ diagonal of readings        = 18

A comparison of the left-hand and right-hand column totals shows the variation caused by glass alone as 30 against 36. These figures indicate that the glass tubes without scratches are somewhat less resistant to breakage than those with scratches, but the difference is relatively small and not significant. A comparison of the top and bottom row totals shows the variation caused by alignment alone as 32 against 34. The difference of the two is again not significant, indicating that alignment is not an important variable. A comparison of the two diagonal totals (the four readings associated with $T_1$ against the four associated with $T_2$) shows the variation caused by tolerances alone as 48 for $T_1$ against 18 for $T_2$. This difference is indeed significant; it indicates that the major cause of failure is too much play (large differential between tube O.D. and yoke I.D.). Attention to this problem of tolerances, which involved very little redesign, eliminated further field failures.

## 9.5     Random and Multiple Balance

The design experiment known as "random and multiple balance" is used if the number of factors or suspected variables is more than four and less than 25. Traditional engineering calculations would, with so many factors, become tedious, prohibitively expensive, and even inaccurate if interactions were involved. Computers could reduce both the tedium and the expense but could not isolate interacting variables. Ordinary statistical methods for investigations involving, say, 20 variables would require $2^{20}$ experiments for a full factorial design and at least 1/64 that number for a 1/64 replicate in a fractional factorial design.

Random and multiple balance, developed first by Dr. Frank Satterthwaite in 1955, is essentially a random sample drawn from a full factorial design. With just 30 tests a broad problem can be scanned and the major factors sorted out. Without delving into the theory of the random and multiple balance design, a simple operating procedure can be described:

1. As many input variables as the design engineer feels are likely to effect a given result or desired output are first selected. Twenty such variables are not uncommon.

2. Two or three levels for each variable are then selected, the range depending upon the extent that each variable is likely to vary under actual operating conditions. The levels, for example, could be labeled "low" or "high" (L or H); or "low," "medium," or "high" (L, M, or H).

3. Next a chart (Table 9.6) is drawn containing a plan for 30 tests, with each test using a combination of different levels for each variable. The level selection is

### Table 9.6 Plan for 30 Tests Using Random Selection of Levels for Each Variable

| Test Run No. | A | B | C | D | E | F | Output: Pellet weight |
|---|---|---|---|---|---|---|---|
| 1 | 0 | 0 | 0 | — | + | — | 56 |
| 2 | — | 0 | + | 0 | + | — | |
| 3 | — | — | + | + | — | — | |
| 4 | + | — | — | — | + | + | |
| 5 | + | 0 | — | + | + | + | |
| 6 | — | + | 0 | — | — | + | |
| 7 | 0 | + | + | 0 | — | + | |
| 8 | — | — | — | 0 | — | — | |
| 9 | 0 | 0 | — | — | + | — | |
| 10 | — | 0 | 0 | + | + | — | |
| 11 | 0 | + | — | + | — | + | |
| 12 | + | + | — | 0 | + | + | |
| 13 | + | + | + | — | — | — | |
| 14 | + | 0 | 0 | — | + | — | |
| 15 | 0 | — | — | 0 | — | + | |
| 16 | 0 | — | 0 | 0 | — | — | |
| 17 | — | — | 0 | + | — | — | |
| 18 | 0 | 0 | + | 0 | + | — | |
| 19 | — | + | + | + | + | — | |
| 20 | + | — | + | 0 | + | + | |
| 21 | — | 0 | — | + | + | + | |
| 22 | + | — | 0 | — | — | + | |
| 23 | 0 | + | 0 | — | + | + | |
| 24 | + | + | 0 | + | — | + | |
| 25 | + | 0 | + | 0 | — | — | |
| 26 | — | + | — | + | + | + | |
| 27 | 0 | — | + | + | — | — | |
| 28 | — | — | + | — | + | — | |
| 29 | 0 | + | — | — | — | — | |
| 30 | + | 0 | — | 0 | — | + | |
| R1 | — | 0 | + | 0 | + | — | |
| R2 | — | 0 | 0 | + | + | — | |

*Levels for Factor or Variable* span columns A–F above.

made at random, using either a random number generator or even the simple toss of a coin. (There are many de-
partures from and refinements to this random balance

to prevent any level of a given variable from being inadequately represented in the experiments.)

The 30 tests represent a sample drawn from an almost infinite number of combinations of levels and variables, 30 being a statistically meaningful number that will yield accurate information about the important and interacting variables. (It is not necessary that 30 separate assemblies be simultaneously made with the combinations of levels and variables shown in the chart. The same general assembly can be built up and torn down after each test.)

*Running the test*     Each of the 30 tests is run using the combination of levels for each variable shown in the chart. The output (weight, voltage, dimension, and so forth) for each test is recorded; sometimes, more than one desirable output may be under observation. A couple of additional tests exactly duplicating earlier combinations (replications) should be thrown in as a check on residual effects.

*Interpreting the results*     Interpretation of the results involves separating the important variables from the unimportant ones and the interacting from the noninteracting, as well as gauging the measure of residual effects (extent of nonrepeatability as revealed in replicated experiments). One quick method of segregation uses Royal McBee edge punch cards:

1. One card is used for each test. The variables (A, B, C, etc.) are set up in columns on the card as shown in Fig. 9.5, and the level for each variable is punched out as determined by the chart (Table 9.6). The result of the test—the output—is recorded on the card.

2. The cards are arranged in order of ascending output (assuming that highest output is the desired result).

3. The geometric pattern of the punches, when the stack of cards is viewed edgewise, tells a great deal about the effects of each variable on the output (Fig. 9.6). Important variables produce a diagonal pattern; unimportant variables show an interleaving pattern; interacting effects tend to show a C (or mirror-C) pattern.

```
o | o     o  | | o     o  | | o     o  | |   o  o  | | o  o  | |   o  o  |  o  o  o  o
o | − o + | | − o + | | − o + | | − o + | | − o + | | − o + |              o
o |   A       |   B       |   C       |   D       |   E       |   F                    o
o                                                                                      o
o                                                                                      o
o        TEST NO. I                              OUTPUT: 56                            o
o                                                                                      o
o                                                                                      o
o                                                                                      o
o                                                                                      o
o                                                                                      o
```

*Fig. 9.5   Royal-McBee card edge-punched for −, 0, or +*

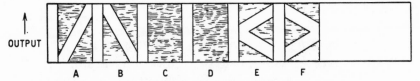

*Fig. 9.6   The pattern when a bunch of such cards are stacked and viewed along an edge*

It must be remembered that this edge-graphing and end-view technique is a rough-and-ready method of segregating variables rather than a refined one. For a more quantitative measurement of main effects, residual effects, and interaction effects, more sophisticated techniques such as scatter plots or matrix analysis must be used.

In Table 9.6 the yield of a process in which six

variables are suspected of being important and interacting is under observation. (Many more variables could have been picked; for the sake of simplicity and clarity, only six were.) The desired characteristic is maximum pellet weight, and the six variables are temperature (A), time (B), pressure (C), cooling rate (D), optional treatment (E), and raw material supplier (F). All have three levels except F, which has only two.

Using a random generator or some other means, a combination of levels of the variables for each of thirty tests is compiled. The coding (+) for high level, (0) for medium level, and (−) for low level is used for all factors except F, in which case the (+) and (–) simply refer to two different suppliers of raw materials. Tests R1 and R2 are replications of tests 2 and 10, respectively, to check for residual effects and experimental accuracy.

The next step is to program the cards as indicated in Fig. 9.5; from Table 9.6 the appropriate level for each variable in Test Run No. 1—(0), (0), (0), (−), (+), (−)—is punched out and the output (pellet weight in grams) entered. The cards for all tests are programmed and arranged in order of ascending pellet weight and the stack viewed edgewise; a configuration like that in Fig. 9.6 results. (The results are idealized; in practice, the demarcation lines are never this clean.) The configuration indicates that variables A and B are responsible for main effects; a high level of A gives higher outputs as does a low level of B. Variables C and D are unimportant; there is no clear trend in output in terms of level. The patterns of variables E and F indicate interaction with other factors. The "culprit" factors that are interacting with E and F can now be pinpointed with matrix analysis. Thus the main and interacting variables have been separated from the less consequential ones, enhancing knowledge of the design, of tolerances to use, and of the factors to control closely.

## 9.6    Design Optimization

Once the most influential factors affecting a design
have been identified (by the techniques described in the
previous section), the investigation can be continued to
determine what levels for these factors will give optimum
output. If the factors are independent (producing only
main effects), each can be moved in the direction needed
for the desired limits to be reached. But if interacting
effects are involved, optimization of the design becomes
far more complicated.

*Evolutionary optimization (EVOP)*    Evolutionary optimization is a technique first introduced by an
English mathematician, G.E.P. Box. EVOP is used to
determine empirically the best combination of levels for
the various factors.

In the case of two factors that interact on any given
output, the technique can be visualized by thinking of
all values of the output as constituting a surface. With
the two interacting factors on the x and y axes, the output
is the corresponding value of z; it is measured in terms
of a percentage of theoretical maximum output. Constant output contour lines (lines along which the altitude, or value of z, is equal) can be plotted against a
range of values for x and y. As an example, assume that
the legibility of marking by a wire marking machine is
dependent on temperature and speed and the interaction between these two parameters. Assume also that
the output can be measured on an arbitrary scale of 100
percent as perfect legibility and zero as total illegibility.
In Fig. 9.7, with temperature and speed on the x and y
axes, respectively, output (legibility) is on the z axis.
Contour lines of equal altitude, therefore, represent a
constant output.

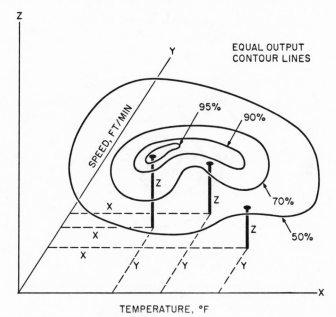

*Fig. 9.7   Evolutionary optimization presentation*

Optimum output is the mountain peak. The topo-
graphy of this output "landscape" is called a response
surface, and the equal output contour lines are lines of
equal response. Any factorial matrix relating a temper-
ature-speed experiment—for example, that of Table 9.7
—can be directly superimposed on the response surface
(Fig. 9.8).

In Table 9.7 the output of 93 percent produced by
combination 2 (medium temperature, medium speed)

**Table 9.7   2 × 2 Factorial Test for Fig. 9.8**

|        | Temperature |         |
|--------|-------------|---------|
| Speed  | Low         | Medium  |
| Medium | 1  70%      | 2  93%  |
| Low    | 3  60%      | 4  90%  |

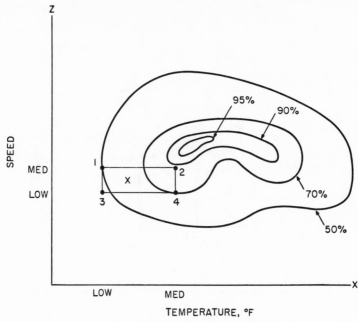

*Fig. 9.8   EVOP approach*

yields the highest contour corner. This gives a clue about the "path of steepest ascent," or the direction in which output will be optimized. If there were no Fig. 9.8 and just Table 9.7, it would still be possible to go "uphill" by moving in a direction at right angles to the intuitively understood lines of equal response.

The grid represented by Table 9.7 could conceivably straddle the peak of the response surface of Fig. 9.8, that is, the additional run is always made at the middle of the grid (point x in Fig. 9.8); this center point of the 2 × 2 factorial is the average value of the factors (temperature and speed). As long as the center response is lower than at least one corner (in Fig. 9.8 the center appears to be at about 60 percent), more 2 × 2 factorial designs must be run, always moving along the path of steepest ascent. When the center is higher than all corners, optimum output has been reached. Figure 9.9

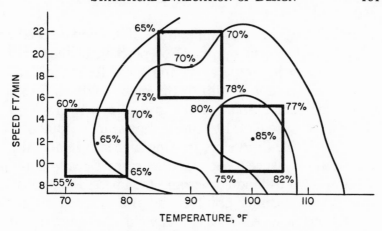

*Fig. 9.9  "Hill climbing" toward optimum value*

**Table 9.8    2 × 2 Factorial Tests for Fig. 9.9**

| Speed, ft/min | Temperature 70° | 80° | Speed ft/min | Temperature 85° | 95° |
|---|---|---|---|---|---|
| 15 | 60% | 70% | 22 | 65% | 70% |
| 9 | 55% | 65% | 16 | 73% | 78% |

| Step 1 | Step 2 |
|---|---|

| Speed, ft/min | Temperature 95° | 100° |
|---|---|---|
| 15 | 80% | 77% |
| 9 | 75% | 82% |

Step 3

is an example of such "hill climbing" toward an optimum output of 85 percent.

The factorial designs for each of the steps in Fig. 9.9 are shown in Table 9.8. In the first step the center point of the grid at a temperature of 75° and a speed of 12 ft/min shows an output of 65 percent. In the second step, moving in temperature and speed from the lowest corner (70° and 9 ft/min) to the highest (80° and 15 ft/min), the center point of the grid at a temperature of 90° and a speed of 19 ft/min shows an output of 70 percent. In the third step, moving higher in temperature but lower in speed because the highest reading is in the lower right-hand corner and the lowest is diagonally opposite, the center point of the grid at a temperature of 100° and a speed of 12 ft/min yields the optimum output of 85 percent.

*Random evalutionary optimization (REVOP)*    The above example involved optimization with two important interacting factors. The same principles can be used with up to four such factors, but visualization of the response surface ceases to be possible. Beyond four interacting factors, it becomes uneconomic to continue EVOP tests because the addition of each factor doubles the number of tests. To keep the total number to a practical, economic level, the technique known as "random evolutionary optimization" (REVOP) is employed.

A nominal set of levels is established for each factor; then random numbers are used to determine the levels for the first test run and the increment and original direction for the succeeding tests. After that, the following guidelines are used:

1. If the first resultant output is *higher* than the nominal (original) output, keep shifting by the same increment in the given direction until there is a decrease in output.

2. If the first resultant output is *lower*, shift in the opposite direction by the same increment. Should this new output be higher, continue as in step 1. Should it be

lower, start again from the original levels with a new randomly chosen increment and direction.

In such "random hill climbing," the optimum output is considered reached if four random tries (eight readings, including opposite directions) are unable to improve the output at a given level of factors.

## 9.7    Overstress Multiple Environment Analysis

Another extremely useful tool in evaluating a design or measuring production and material conformance to a proven design is the technique of overstress multiple-environment analysis. Much conventional product testing is "success testing"—testing to no failure under room ambient or one at-a-time environment conditions. The fewer the failures, the better the design. The strategy of overstress testing, on the other hand, is to induce failures deliberately, through overstress (under simultaneous multiple environments, sequential ones, or both), and thus learn about the adequacy of design over the equipment's useful operating life. By this method, weak links of design that might not show up normally until large quantities of the product are in the field can be eliminated and corrected before it is too late.

*Operating rectangle*    Conventional single-environment testing cannot reliably predict how a product will withstand stress caused by the *interaction* of the stresses for which it tests separately. However, these single-environments—temperature, shock, vibration, humidity, transients, electrical and mechanical load—can be combined into a single stress scale made up of increments of each environment. On such a stress scale, two significant points can be located: one at which all stresses are at a maximum specified level simultaneously —called *design stress* or *rated load*—and a second and higher one beyond which it becomes impractical to apply further stress—called *maximum practical overstress*.

With respect to operating life, there is also a scale, with one important point—service life—up to which the product is expected to perform without failure. By combining these two scales—stress and life—an operating area can be defined (Fig. 9.10). The object of design should then be to clear this area, known as the operating rectangle, without failures.

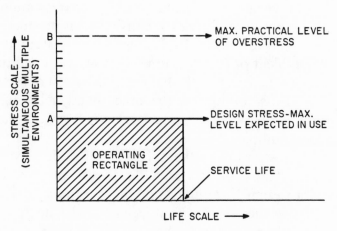

*Fig. 9.10   Plotting the stress-life operating area*

*Stress path*     The first step in determining the overstress path is to establish the maximum level of stress to be expected in actual use by taking into consideration all predictable environments and their degree of severity (level A in Fig. 9.10). The next step is to determine the maximum practical level of stress contributed by all environments that the equipment can withstand without falling apart (level B). The distance between levels A and B is then arbitrarily divided into ten increments of overstress.

The critical stress path is shown in Fig. 9.11, a configuration depicting a three-factor stress environment. The product is progressively stressed along this predetermined path until the critical corner (level A of Fig. 9.10) is reached. If the design is reasonably good, there should be no failures up to this point. The stress is then increased by the predetermined increments until a failure occurs. The cause is determined, the correction

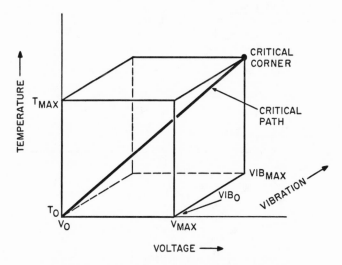

*Fig. 9.11    Three-dimensional stress-life plot*

made, and the overstress continued until the next failure occurs. It is desirable to accumulate at least four or five failures for each failure mode. For practical purposes, it can be assumed that all occur at time zero.

It is not necessary to test a large number of units for the tests to be statistically meaningful. Generally, seven units are preferred. But if costs or time preclude this number, even two units may be adequate. What is necessary is that failures be induced. Without failures, no knowledge of inherent weaknesses is possible.

Figure 9.12 indicates the distribution of failures caused by a single failure mode on the stress scale. Assuming that all these individual failures have cleared the design stress level, a statistical K-Factor analysis or a Weibull plot can be made to determine the tail-end of the failure distribution.[1] If this population tail-end still

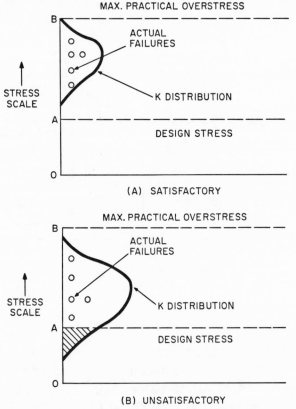

*Fig. 9.12    Failure distribution on stress scale*

[1]    The reader is referred to standard texts on statistics and reliability for detailed treatments of K-Factor analysis and Weibull plots.

clears the design stress level (Fig. 9.12A), no corrective action need be taken; if not (Fig. 9.12B), back to the drawing board!

As an example, an automobile radio is to be designed to withstand a maximum temperature of 130°F, a maximum vibration of 5 g and a maximum battery voltage of 15 V. Testing the product up to the maximum stress of each of these three environments in isolation is not enough because it may not bring out the weak links in the design. But subjecting a few radios to stresses of temperature, vibration, and battery voltage simultaneously is a different matter. Raising the temperature in steps from the design stress level of 130°F to, say, 180°F; raising vibrations from 5 g in similar steps to, say, 15 g; and raising battery voltage from 15 V in similar steps to 25 V may deliberately induce failures which can then be examined in detail.

Thus, the $T_0$ of Fig. 9.11 becomes 130°F and $T_{max}$, 180°F; $V_0$ becomes 15 V and $V_{max}$, 25 V; $Vib_0$ becomes 5 g and $Vib_{max}$, 15 g. As temperature, voltage, and vibration are simultaneously increased on a preproduction experimental quantity of 10 radios, failures are noted. Specific failures can be corrected by replacement or repair and the radio returned to continue the test.

If there are only isolated failures of varying kinds above the design stress level, it can be assumed that the design is adequate. On the other hand, if there are four or more failures of a single type (failure mode), the pattern of such failures should be plotted on the stress scale of Fig. 9.12. It is often sufficient to observe the pattern of this sample of four or more failures visually to determine the tail of a distribution of a whole population of failures in the same mode. If the tail dips below the design stress level, the particular failure mode must be thoroughly investigated and corrected.

In Fig. 9.12B, five failures of a single type are shown,

two of them occurring at an identical stress level. Although all five failures occurred above the maximum design stress level, some were so close to that level, including the two occurring at the same level, that it is conceptually apparent and statistically verifiable that failures will also occur below the design stress given a large enough number of production units. In Fig. 9.12A, failures occur at sufficiently high and favorably distributed levels to make failures below design stress very improbable.

*Overstress life testing*      Testing along the stress scale reveals the most likely failure modes at time zero. The next step is to confirm design adequacy with time. Some failure modes degrade with time; others do not. Obviously, the high costs of prolonged testing to the end of service life can become a problem here. In equipment of a cyclical character, such as relays, switches, and the like, the time span can be shortened by constant cycling. In continuously operated products, however, the time span cannot be artificially shortened. The strategy in such cases is extrapolation. Failure distribution at time zero is compared with failure distribution at two or three time intervals considerably short of service life-time. These distributions would then be extrapolated to see if they still clear the operating rectangle at service life (Fig. 9.13). This approach does not automatically give any assurance of protection throughout service life but is a reasonable prognosticating device.

*"High Time" experiments*      Sometimes, customers' equipment that has been in actual operation in the field can be brought back and subjected to further multiple-environment overstress analyses. The advantage of such tests is that these units have already accumulated

aging time in the field—usually with no cost or delivery tie-up problems for the company. If there are adequate field records of operational time, it may be possible to commence an overstress analysis on these units at time

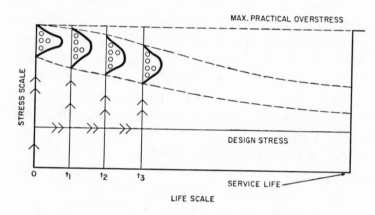

Fig. 9.13    Extrapolation to service life of part

intervals well beyond $t_3$ in Fig. 9.13. Thus the ability of the units to clear design stress at full service life could be more positively evaluated and costly field failures anticipated before they actually occur.

## 9.8    Value Engineering In Quality Control

As a management tool, value engineering has its principal usefulness in the elimination of excess costs without the deterioration of required quality. Quality control, another management tool, coordinates various disciplines to assure full customer satisfaction at the most economical levels. It is clear that value engineering and quality control have similar objectives. These objectives are not incompatible; reducing cost and improving quality can complement one another. Hence there is a natural tie-in between value engineering and quality control, and their principles can be harnessed together to reduce quality costs.

Among the many applications of value engineering to quality control are the challenging of specifications and requirements; the calculation of a cost-performance profile; the establishment of realistic outgoing quality levels, incoming quality levels, and in-process quality levels along operations research lines; statistical tolerancing to ease component tolerances; pre-control techniques to provide economical insurance against poor quality; more economical sampling plans; and the measurement, control, and reduction of quality costs.

*Establishing quality levels*    One of the prime tasks of quality control, in cooperation with top management, is to set out-going quality levels. These are often arbitrary decisions, for example, 1 percent de-defective at the shipping door. A value engineering approach should attempt to quantify the costs for various percentage levels of defectives. On the one hand, there should be estimates of warranty, rework, retrofit, and complaint follow-up costs as well as estimates of the loss of sales because of customer dissatisfaction. Balancing these external costs, there should be an estimate of the internal costs of attaining the quality level desired, in terms of appraisal, internal failure, and preventive costs. The percentage level of defectives that yields the lowest overall costs should be selected as the proper outgoing level.

It is not easy, of course, to arrive at such costs. But the attempt to quantify the various elements necessitates the establishment of quality levels for incoming materials. In many companies, where numerous parts and components and raw materials are procured from outside vendors, it is customary to specify quality levels in terms of percentages of acceptable quality level, AQL, and then use sampling plans that check parts to these specified AQL's. The acceptable quality level is defined as that percentage level of defective units that, on an average, will be accepted 95 percent of the time through the use of the given sampling plan. An AQL of 1 percent indicates that if a lot containing 1 percent defective is submitted to the sampling plan, that lot would have a 95 percent chance of being accepted. The AQL's determine the appropriate sample sizes.

Unfortunately, most AQL's are stereotyped values, established by custom and precedence, rather than by intelligent analysis. The value engineering approach, however, manages to evaluate the costs of such inspection against the costs that would be incurred further downstream if no inspection were performed at the incoming stage.

J.W. Enell has evolved a simple means of selecting the most economical AQL's (based on the MIL-STD-105D sampling plan. He defines the break-even point, $P_b$, as

$$P_b = I/A$$

in which:

I = cost to inspect one unit
A = damage done when one defective unit slips through inspection (the cost of replacing or repairing the defective unit itself is assumed to be small in comparison with the total damage done)

Based on this concept, Table 9.9 shows the recommended AQL's for various ratios of unit inspection cost to unit damage.

Table 9.9    Recommended AQL's for Various $P_b$ Ratios

| $P_b$ | Recommended AQL. percent |
|---|---|
| 1:900 | 0.015 |
| 1:400 | 0.035 |
| 1:300 | 0.065 |
| 1:200 | 0.10 |
| 1:150 | 0.15 |
| 1:90 | 0.25 |
| 1:65 | 0.40 |
| 1:50 | 0.40-0.65* |
| 1:33 | 0.65-1.0* |
| 1:25 | 1.0-1.5* |
| 1:20 | 1.5-2.5* |
| 1:12 | 2.5-4.0* |
| 1:9 | 4.0-6.5* |

\*   Use the smaller AQL's for lot sizes less than 1,000

The strategy, then, is to determine the ratio of the inspection cost to damage cost for a given unit and then obtain the appropriate AQL. This approach almost always results in a higher AQL than is usually allowed by the quality control department of most companies. This, in turn, means a smaller sample size or fewer needless rejections, or both.

Even greater savings can be made if the quality of incoming supplies can be predicted. If this quality—as predicted by past history, supplier reputation, and knowledge of the fabrication process and state of the art— is estimated to be much better than the required AQL, no inspection is needed. If the quality level is predicted to be much worse than the required AQL, a complete inspection will be the most economical approach. If the quality level is at neither of these extremes or is of a fluctuating, variable nature, sampling is the answer. The several sampling plans which can be used are described in a later section. It should be noted that application of techniques such as these reduce incoming inspection costs by 25 to 50 percent.

*Selection of vendors*        One of the frequent frustrations of quality control personnel is their inability to convince management of the desirability of terminating a vendor with a proven record of bad quality. Often, for example, the purchasing department may champion the continuation of such a vendor because of price, cooperativeness, or just habit. Needless arguments can be avoided with a value engineering attack on the problem. Management and purchasing cannot be influenced effectively by percentage statistics. But if the *total* costs associated with a particular vendor can be quantified and compared with those of another vendor, a stronger dollars-and-cents case can be established by quality control and effective action result.

Such cost comparisons between the two vendors should include the following parameters:

1. Vendor price
2. Costs of incoming inspection, including sample and complete inspections
3. Costs of quality control engineering investigations for accept–reject decision
4. Costs of corrective action follow-up (paperwork, visits to vendors, and the like)
5. Costs of return of material to vendor (material handling, return orders, and the like)
6. Costs of reinspection
7. Costs of delays in production caused by high rejections.

It is not necessary, of course, to determine such costs for all vendors or even for those who are a recurring problem. The approach is applicable to only those vendors who present the greatest cost problem. Furthermore, attempts to ascertain whether one's own company (rather than the vendor) has its house in order should precede such investigations. But, in the final analysis, an objective, quantitative dollars-and-cents comparison affords the most positive guarantee of getting one's money's worth.

To be completely fair to the purchasing department and to look at the total quantitative picture, factors such as timeliness of delivery, vendor cooperativeness and flexibility, and the impact of such matters on other divisions of the company should also be considered. Some of these may indeed be difficult to quantify, but an attempt should be made. Once numerical estimates are presented they can be challenged; the result of such challenges is nearly always a refinement of the figures and their eventual acceptance by all parties concerned.

## 9.9   Statistical Tolerancing

Statistical tolerancing is an extremely valuable engineering tool for bringing about outstanding cost reductions by opening up unnecessarily close tolerances. This economy can be achieved with very little risk of degrading quality.

The traditional method of tolerancing has been "additive" tolerancing, in which tolerances of the component parts of an assembly are selected so as to add up to equal, but never exceed, the assembly tolerance. In "chance" tolerancing, the sum of these tolerances is allowed to exceed the assembly tolerance as a matter of pure luck. "Statistical" tolerancing, on the other hand, is based on the familiar normal curve (Fig. 9.14).

Tolerances for most parts produced from machines or processes that are *under control* have a frequency

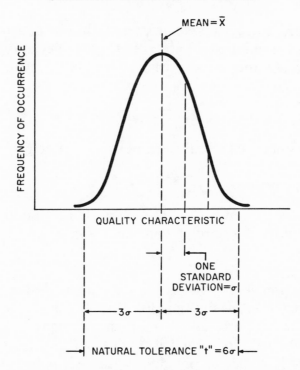

*Fig. 9.14   Normal curve of statistical tolerancing*

distribution that approximates a normal curve. The tails (or most deviant parts) of this curve contain extremely few of all the parts produced (0.3 percent of the total parts population at the 3σ limit, where σ equals one standard deviation; in Fig. 9.14 the natural tolerance band of the process is seen to be 6σ, or 3σ on either side of the numerical mean, x).

When the tolerances for all parts are added, statistical theory states that the resulting distribution of the assembly tolerance is also a normal curve but with a variance (square of the standard deviation, designated by the symbol $\sigma^2$) that is equal to the sum of the component variances, that is:

$$\sigma_{x_A}^2 = \sigma_{x_1}^2 + \sigma_{x_2}^2 + \sigma_{x_3}^2 + \dots$$

in which $x_1$, $x_2$, $x_3$, etc., represent the component parts of an assembly $x_A$. Since the standard deviation, $\sigma$, is the square root of the variance, the above equation can be rewritten as

$$\sigma_{x_A} = \sqrt{\sigma_{x_1}^2 + \sigma_{x_2}^2 + \sigma_{x_3}^2 + \dots}$$

Also since the natural tolerance t is equal to $6\sigma$,

$$t_{x_A} = \sqrt{t_{x_1}^2 + t_{x_2}^2 + t_{x_3}^2 + \dots}$$

This means that natural tolerances do not add up arithmetically but according to a root mean square (R.M.S.) law.

*Example*     An assembly has a specified tolerance of $\pm 0.008$ in., as shown in Fig. 9.15. It is made up of four components, each equal and each with equal tolerances (for the sake of simplicity). The additive tolerancing method would simply divide the assembly tolerance equally among the four component parts at $\pm 0.002$ in. for each component.

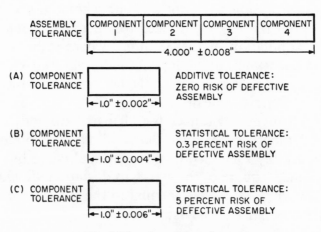

Fig. 9.15    Assembly tolerance

Using statistical tolerancing, however, and again assuming equal tolerances, we obtain:

$$t_{x_A} = \sqrt{4t_{x_1}{}^2}$$
$$2t_{x_1} = t_{x_A}$$
$$t_{x_1} = t_{x_A}/2$$
$$t_{x_1} = \pm\,0.008/2 = \pm\,0.004 \text{ in.}$$

This statistical component tolerance of $\pm\,0.004$ in. is thus twice the additive tolerance of $\pm\,0.002$ in. Even more important, this opening up would mean only a 0.3 percent risk of creating defective assemblies since the assembly tolerance would fall within the $\pm\,3\sigma$ limits of Fig. 9.14. From statistical theory, it is known that 99.7 percent of all readings in a normal curve fall within these limits.

If this risk of 0.3 percent can then be opened up to, say, 5 percent, the component tolerances can be opened up even more. This can be accomplished by having the assembly tolerance fall within the $\pm\,2\sigma$ limits since statistical theory informs us that these limits encompass 95 percent of all readings, leaving a mere 5 percent fallout. In this case,

$$4\sigma_{x_1} = 0.008$$
$$\sigma_{x_1} = 0.002$$

And since $\sigma_{x_A} = 6\sigma_{x_1}$,

$$\sigma_{x_A} = 6 \times 0.002 = 0.012$$

We know from before that $t_{x_1} = t_{x_A}/2$, and since $t_{x_A} = \sigma_{x_A}$,

$$t_{x_1} = 0.012/2 = 0.006$$

Thus the natural component tolerance is $\pm\,0.006$ in., indicating that the component tolerances using 95 percent statistical tolerancing are *three times* those obtained by using additive tolerances.

*Modifications of theoretical calculations*    The above statistical theory is based on two assumptions representing idealized production. It assumes first, that the grand average of each component falls precisely at the nominal value specified, and second, that the frequency distribution for each component is normal. Neither of these assumptions is always valid. To account for these nonnormal conditions, the R.M.S. law can be "tightened" somewhat by the expression:

$$t_{x_A} = 1.5\sqrt{t_{x_1}^2 + t_{x_2}^2 + t_{x_3}^2 + \dots}$$

in which the factor 1.5 is a safety factor against nonnormality. Using this formula in the previous example, we obtain

$$t_{x_A} = 1.5\sqrt{4t_{x_1}^2}$$
$$= (1.5)(2)(t_{x_1}) = 3t_{x_1}$$
$$t_{x_1} = t_{x_A}/3$$

In the case of the $\pm 3\sigma$ limits this becomes

$$t_{x_1} = 0.008/3 = 0.00266$$

In other words, component tolerances have been opened up 33 percent from the additive tolerance (0.002) but not 100 percent as in the idealized condition.

In practice, it is recommended that the first cautious step be to use the 1.5 safety factor and then reduce it gradually towards unity as actual experience and the control over various processes allow.

*Nomograms*    Nomograms can be used to simplify calculations of component or assembly tolerances. Figure 9.16 is an example. The two component tolerances are entered on the side scales and connected by a straight-edge. The intersection on the center scale gives the assembly tolerance.

For more than two components, the "subassembly"

tolerance reading calculated for the first two components is entered on one side scale and the third component tolerance on the opposite side scale. The tolerance for the three components is then read on the center scale. This process is repeated for all components.

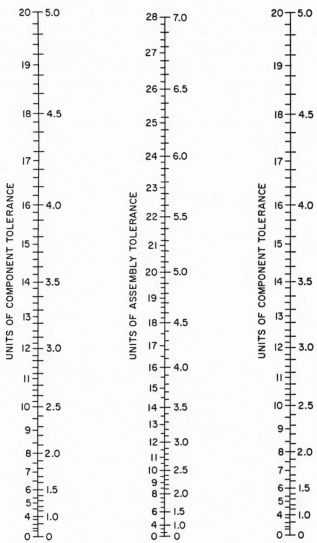

*Fig. 9.16   Nomogram: Result of two tolerances*

To go from an assembly tolerance to desired component tolerances, the assembly tolerance is entered on the center scale. This point becomes the pivot for the straightedge, and any combination of two component tolerances can be read off on the side scales. For more than two components, the same procedure is used, but the pivot will then give combinations of side readings, one of which will be the tolerance for the first component and the other a combined tolerance for all the other components. This combined tolerance can now be entered on the center scale to produce two more readings for the second component and a combined tolerance for the remaining components. And so on.

For simplicity, Fig. 9.17 can be used as a first approximation. Combined tolerances for two or more components may be determined, but only if the component tolerances are equal.

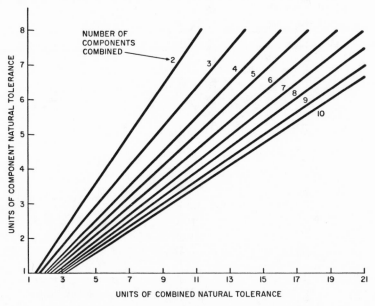

*Fig. 9.17   Chart for combined equal tolerances*

*Practical applications*          Statistical tolerancing has wide application in industry in mass production as well as in job shop production. Its use in mechanical assemblies is straightforward because the mathematical relationship between the assembly and its component parts is linear.

Statistical tolerancing is also useful in separating process error from error caused by gaging and by the operator or inspector. The distribution-of-tolerance equation is expressed as

$$\sigma_t^2 = \sigma_p^2 + \sigma_g^2 + \sigma_o^2$$

in which:

$\sigma_t^2$ = total variance

$\sigma_p^2$ = variance caused by the process alone

$\sigma_g^2$ = variance caused by the gage or measuring equipment alone

$\sigma_o^2$ = variance caused by the operator or inspector alone

This equation is particularly useful if nominal values add algebraically, as in electrical work for the resistance of several series resistors, the gain of multistage amplifiers, and the like.

# 9.10    Precontrol

Another value engineering application in controlling process quality is to substitute precontrol for control charts. Control charts were introduced as a periodic check on processing to keep defective products at the lowest possible level. Although they can be costly in terms of frequent inspection and paperwork, they are not so costly as making no checks whatsoever.

Control charts measure the "pulse" of a particular parameter in a process and determine when corrective action becomes necessary. Typical control chart techniques consist of taking a sample of four or five readings periodically (every hour on the hour, for example), calculating the average (A) and the range (R) of these readings, and then plotting them on charts. If the readings fall within predetermined limits, the process is under control. If the readings are outside these limits, the process must be stopped and corrected before production can continue.

Precontrol provides the basic function of a control chart more quickly, more cheaply, and usually with less statistical error. It is an insurance type of check, an early-warning system. It has the following advantages over alternative plans:

1. It is lower in cost than almost all other control plans for comparable quality.

2. It serves as a set-up plan, starting with the first unit produced, and hence is ideal for short production runs (and job shops).

3. It automatically adjusts inspection frequency to an economic level and so is ideal for long production runs.

4. It indicates drifts in process error and is thus a good diagnostic tool (a "variables plan") as well as a qualitative one (an "attributes plan").

5. It assures that the percent defective of the product will not exceed a specified level.

6. It requires little or no paperwork.

7. It is easy for shop operators, inspectors, and supervisors to understand and use.

8. It works directly from specification limits; other limits are not needed.

9. It quickly determines if the "material" tolerances

of the process are too tight or too loose compared to specification tolerances.

Figure 9.18 shows the relationship between specification tolerances or limits and the precontrol lines. The specification tolerance is cut in half to determine the precontrol limits, which are then centrally located equidistant from each specification limit. If precontrol is not used in quality control work, it is usually necessary

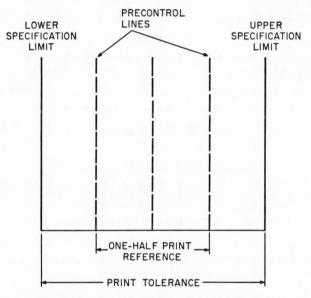

Fig. 9.18    Precontrol lines for given limits

to run a process capability study involving 25 to 50 units at the start of a given process to determine the worth of the process. With precontrol, only 12 consecutive units must fall within the pre-control limits for the process to be considered certified for larger production. (From this point on, normal control chart techniques are used only as backup safety insurance.)

At the start of a given day of production, two units

consecutively produced by the process are measured. If both or even one of the readings fall within the precontrol limits, the process can be certified as good for the next period of time (say an hour or even a day, depending upon process stability, economics of rejection, and previous history). No further checks are necessary, then, till the next period of time. If, however, both the readings fall outside of the precontrol limits, yet on the same side and still within the specification limits, it can be assumed that the process is drifting and should be stopped to make an adjustment or a setting. If both the readings fall outside the precontrol limits but on opposite sides of these limits, it can be assumed that the process is malfunctioning because of faulty material, machine difficulty, or operator or equipment error. Again the process should be stopped to determine the cause of the trouble and to correct it. After correction, five consecutive units must fall within the precontrol limits before full production can be resumed.

It can be seen, now, why precontrol has tended to make the old control chart obsolete. Samples are reduced by more than one-half; their frequency is decreased from, say, eight times a day to once a day; no calculations are needed to compute averages, ranges, and the like; and finally, the trend towards defective quality is apprehended much sooner.

## 9.11    Sampling Plans

Value engineering techniques can also be applied to a wide assortment of appraisal functions. The possibility of eliminating several checks that may have been inaugurated early in the life of a product to reassure a design engineer but that have outlived their usefulness is always a fertile field for value engineering efforts. The possibility of changing from the one-hundred-percent inspection dictated by habit to an in-process sampling plan is another. The substitution of a particular sampling plan more economical than, say, a traditional military standard is a third. The substitution of a spot-check for a sampling plan is yet a fourth.

The value engineer must become familiar with the wide assortment of sampling plans that are available, their salient features, and where they can be applied most economically. There is a pronounced tendency in industry, because of the lack of such familiarity, to apply a ready-made plan like Military Standard 105D to all situations regardless of economics. Table 9.10 is a capsule summary of the most useful sampling plans along with their specific areas of application. It is impossible, of course, to discuss all the details of each of the plans. Numerous government and industry publications give

**Table 9.10  Salient Features of Several Sampling Plans**

| Plan | Application | Salient Features |
|---|---|---|
| Mil.-Std. 105 D (Attributes) | General application in receipt inspection; end items; in-process inspection; supplies; paperwork and procedure | Slanted towards AQL's and thus favors "producer"; reduces rejection risk of good lots; poor insurance against bad lots; provides a simple, useful starting point in sampling plans |
| Dodge-Romig (Attributes) | General application, particularly where rejected lots can be 100 percent inspected | Weighted towards L.T. P.D. and A.O.Q.L. and hence favors "consumer", but inspection costs are high |
| H-107 (Attributes) Single-level continuous | Suitable for high volume continuous production | Penalizes poor quality; rewards good quality |

## Table 9.10 Several Sampling Plans (Cont'd)

| | | |
|---|---|---|
| H-106 (Attributes) Multilevel continuous | Suitable for high volume continuous production | Same as H-107; it is the most economic of all plans if the quality level is better than specified |
| Chain sampling (Attributes) | Useful when costly or destructive tests are called for | Gives a second chance in C = 0 plans by going back in history to include more units in sample if a single defect is found in present sample |
| Discovery sampling (Attributes) | General application where probabilities of defective lots can be estimated | Can be fitted into any other attributes plan with the object of reducing sample sizes based on estimated probabilities of lots being defective |
| "Identity inspection" (Attributes) | For lot-by-lot acceptance where previous quality history indicates no defects | Lots treated on a modified continuous sampling plan; no checks other than "identity inspection" for four out of five lots after "proving in" period |
| "Identity inspection" by characteristics (Attributes) | For lot-by-lot acceptance where lots are few, quality is uneven, and knowledge of process is a clue to amount of checking required | Every lot is inspected using a sampling plan; but only important or failed characteristics are checked on each sample; the remaining characteristics are checked on a modified continuous sampling plan using only "identity inspection" |
| Mil.-Std. 414 (Variables) | Especially useful when only a few characteristics need be controlled; distribution must be normal or near normal | Sample size much smaller than attributes plans with the same AQL's; calculations involved |
| Shainin lot-plot (Variables) | Same as Mil.-Std. 414 comments; particularly useful where large lots are involved | Requires only 50 pieces to be measured, regardless of lot size; calculates Z defective in lot more accurately than any other plan; requires some calculations |

detailed instructions on their use. But a few examples will serve to illustrate how Table 9.10 can be utilized to improve quality protection and simultaneously reduce costs.

*Example 1*     An ordnance manufacturer has been using Mil. Std. 105D to check a particular part that he receives in his incoming inspection department in lots of 100,000 items. The part is relatively simple, with only three or four quality characteristics to be checked, of which only one is critical. The quality levels from the supplier have been erratic. What is the best plan to use?

Since the lot sizes are large and the number of important quality parameters are few, a variables sampling plan (involving measurements) rather than an attributes sampling plan (good versus bad) is in order. Of the variables sampling plans, the Shainin lot-plot plan would offer the greatest economies, since the sample size will always be limited to 50 items (as opposed to the 500 to 700 items required by Mil. Std. 105D). The calculations can be reduced and streamlined so that shop people can use it.

*Example 2*     An assembly line has been using Mil. Std. 105D to control the quality of its outgoing product. The production rate is about 1,000 units per day. The quality record is good, but it is both difficult and expensive to accumulate lots to apply Mil. Std. 105D sampling. What is the most effective plan to use?

The key here is that the current quality level is acceptable—perhaps equal to or better than the level specified. This fact points to a continuous sampling plan such as H-107 or H-106. These can monitor a continuous stream of products without having to wait for lot-by-lot accumulation. The production rate is also large enough to sustain a continuous sampling plan. If administered properly, and if the production quality level continues to be good, such a plan can cut the sample size of Mil. Std.105D to about one-fifth.

*Example 3*     A vendor has been submitting a certain item to a customer to the complete satisfaction of the latter's incoming inspection department. Several lots have been accepted, and no lot has been returned to the vendor following the customer's Mil. Std. 105D sampling. The item itself is simple and easily replaced in case of failure in downstream production. What is the most economical plan to use?

The evidence clearly calls for "identity inspection." Since the previous history indicates no failures and since a rejection can easily be detected and corrected inexpensively further downstream, it is safe to sample-inspect only every fourth or fifth lot, with inspection on every lot being confined to just a visual check of one item from the lot to establish proper identity. This procedure would reduce the total inspection on the item by 50 to 80 percent as compared to that required by Mil. Std. 105D.

*Example 4*     A pharmaceutical company has been using Mil. Std. 105D to inspect the chemical composition of drums of incoming powders from a given supplier. There are usually 80 to 100 drums in each lot. The quality has, in the past, tended to vary from one lot to the next. What is the optimum sampling that the company should perform on the incoming drums?

This problem presents a slight variation on identity inspection. A principal reason for any randomization in sampling is to assure that every item or portion of an item in a lot has a equal opportunity of being selected for the sample. If there is evidence in the above example that there is considerable difference in quality or chemical composition between one drum and another in the same lot, or even differences within each drum, a complex randomization procedure must be used to select the sample. On the other hand, in some instances there could be considerable homogeneity within each lot so that there will be little variation from one drum to the next and from one section of a drum to another. In such cases, the sample size could easily be reduced to one drum to establish the quality level for the entire lot of 80 to 100 drums.

# 10

# *Organizing a Value Engineering Program*

When company management first considers the use of value engineering, it is faced with a maze of questions that must be solved if the program is to be successful. The most important questions are:

1. Can value engineering be profitably used by the company?
2. In what areas should it be used?
3. Who should be responsible for it?
4. How many people will be required, and at what level should they report?
5. How much will the program cost, and what return may be expected from it?

The answer to these questions will, of course, vary greatly according to the size and character of the organization. A number of factors, some of which may be unique to a given organization, must be considered in designing a program. The proper evaluation of these factors, coupled with a basic knowledge of what value engineering is and how it functions, will result in a well-designed program. The factors can be listed as follows:

1. Is the organization a product- or service-type organization?
2. What is its size?
3. What is its existing organizational structure?
4. What is its product cost content?

*Type of organization*    Value engineering was initially used in product-oriented organizations, and in this area has had its greatest application and growth. That is not to say that it cannot be applied successfully in service-type organizations; numerous examples substantiate its successful application to these. But, since the bulk of experience to date has been in product-type organizations, the suggestions offered in this chapter will be directed towards that group.

*Size of organization*    The size of an enterprise determines the number of value engineers to be employed; whether the effort is to be a centralized or decentralized operation; and what the reporting level will be.

*Organizational structure*    Value engineering must be fitted into the existing organization with the least possible disruption and without sacrificing any of its effectiveness. The objective should be to make sure that it receives the best top management support available since this support is absolutely necessary to a successful program. The type of overall organizational structure will in large measure determine whether it should be line or staff, whether centralized or decentralized, and whether located in engineering, purchasing, manufacturing, or elsewhere.

*Product cost content*     The extent to which raw materials, purchased components, engineering, efforts and manufacturing procedures affect the total cost of a product will also influence the organizational location of the value engineering activity and the extent to which value engineering should be applied in these various areas. Obviously, a product requiring a high degree of sophisticated engineering should receive a substantial VE input in engineering, at the conceptual design and the product design phases. On the other hand a high-volume product of rather simple design should be studied intensively in the manufacturing and purchasing phases. In general, the value effort should be devoted initially to areas that account for the greatest expenditure of dollars since these are the areas where substantial savings are most likely to be made.

The following is a list of the planning and organizational steps for setting up a value program:

1. Research in value engineering philosophy and techniques
2. Review of value engineering functions, responsibility, authority, and content and method of operation
3. Analysis of specific factors within a company that affect value engineering
4. Indoctrination of top and middle management
5. Selection of director for the value program
6. Selection of consultants
7. Selection of value personnel
8. Program initiation
9. Personnel training
10. Program operation
11. Measuring effectiveness and reporting
12. Implementation

# 10.1    Research in Value Engineering

Before much progress can be made in planning for a value program, the would-be user must gain a basic knowledge of what value engineering is, for which the earlier chapters of this book provide the necessary information in depth. But a knowledge of value engineering without any experience in its application is much like the knowledge of skiing gained by reading a book on the subject without ever setting foot on the slopes; one can gain an intellectual understanding of the sport but not be able to get safely to the bottom of a hill. Appreciation of the satisfaction that mastery of the sport can give requires experiential knowledge.

It is earnestly recommended that some value engineering experience be acquired through the application of techniques on live hardware. This applies particularly to the manager responsible for this activity, who most needs an imaginative understanding of the full potential of this management tool.

# 10.2    Method of Operation

Broadly speaking, it is the function of value engineering to improve profits by improving the value of the company's products or services. This task is fulfilled by:

1. Training personnel in the knowledge and use of value engineering techniques and promoting the application of such techniques
2. Providing value engineering services to Engineering during design and development; this involves

working directly with design groups, attending design reviews, and setting cost targets for designs

3. Organizing and directing value engineering task force groups to improve existing designs or assisting Engineering in meeting cost targets on new designs

4. Organizing and directing the application of value engineering techniques in purchasing, manufacturing, and other company functions

5. Accumulating and disseminating cost information necessary for value studies

6. Accumulating and disseminating information on special products and specialty vendors

7. Establishing policies, controls, and reporting procedures for the value program.

Since, as is obvious, the value engineering activity must work across many corporate functions, it requires a substantial amount of people-contact and salesmanship. Furthermore, sufficient authority or appeal to authority must be incorporated into its organizational design to provide for value engineering's proper input and application throughout company operations and to insure the effective implementation of qualified value engineering proposals resulting from studies made by either the value engineering department or value engineering task force teams. Otherwise, instead of being useful, the work of the value engineering group will frequently be bypassed under the normal pressure of business, and the group will cease to function effectively.

## 10.3    Size and Physical Location

Whether to have a centralized or decentralized value engineering activity will depend to a large extent on the size of the company and its physical location. A large corporation with a number of plants located throughout the country—or throughout the world—would find a decentralized activity more effective. Each plant could have its own activity, reporting to its own management, while the value engineering program as a whole was directed from corporate headquarters. The corporate activity would function as a coordinating and training organization, providing assistance to the plants in selecting projects, training personnel in value engineering techniques, conducting value engineering workshops and special studies, and establishing goals for the value engineering activity. It would also monitor each plant's value engineering programs, would report to corporate management on the entire program, and would indicate areas for improvement.

For the large corporation located at one plant site, the question of making value engineering a corporate or divisional activity becomes more difficult to answer. Factors such as the existing organizational structure and particular personalities may play a more important role in making such a decision than the company's size and location. Centralization has the advantage of giving the value engineering activity its own identity as a recognized corporate function and—bringing together the talents of several different individuals—making a competent team capable of solving value problems while operating across all company functions.

The greatest danger with a centralized activity is that it may lose contact with the mainstream of divisional happenings and become isolated and therefore relatively ineffectual. On the other hand, while decentralizing the activity has the advantage of making sure the value engineer is "where the action is," it too has its weakness; the value engineer's time and effort may become diluted with assignments of line operational problems rather than value engineering problems.

The size of a company will obviously affect the size of its value engineering operation, although not to the extent that it is the limiting or determining factor. A company with sales between two and five million dollars should definitely consider assigning one individual full time to the value engineering function. The number of value engineers required does not necessarily increase in proportion to higher sales volume, since other factors (such as product complexity, variety, and volume) will have their effect on group size. A company engaged in the design, development, and manufacture of highly engineered products (and therefore employing a large staff of engineers) may need a larger value engineering service group than a company which is predominantly a production operation manufacturing only a few stand-ard products on high-volume equipment. The best in-formation available at this time indicates a range of from five to ten value engineers as the average for most large engineering groups. A rough rule of thumb to follow in deciding group size is one value engineer for every fifteen to twenty engineers in a company. In the final analysis, the most efficient size for a value engi-neering activity will be determined by experience.

The character of the operations and products of a company may help determine the emphasis of the value engineering program. If highly engineered products are being produced in limited quantities and in conformity with rigid specifications (as in many aerospace com-panies), a value engineering effort concentrated at the conceptual-design and product-design levels would be most wothwhile. Once such products have reached the production stage and have been tested, qualified, and tooled, it is usually not financially feasible to make significant design or manufacturing changes. Converse-ly, if products are not overly complex and are made in high volume, considerable value engineering effort can be devoted to them even after release and initial production. With such high-volume items, the possible improvements in design and production usually warrant a value engineering effort at this latter stage because of the substantial dollar volume of the items being produced. Often value improvements become possible because of new materials, processes, and technology not available at the time of the initial design. It should always be borne in mind as a general rule, though, that the greatest impact value engineering can have is during the concept and design phase.

Another product characteristic that will affect the type of program designed is the extent to which purchased components determine total product cost. If end items consist primarily of assembled purchased items, a significant value engineering effort should be directed toward the purchasing function. This will include activities such as investigating the use of alternative items to perform a given function; working with vendors to reduce their costs (in effect, their prices) by reviewing with them functional requirements and manufacturing costs to eliminate nonessentials; and providing vendors with motivation and direction in the use of value engineering techniques. Companies heavily weighted with purchasing functions should consider locating their value engineering effort in the purchasing department or otherwise provide for sufficient coverage of its activity in their value engineering organization.

Intangible factors, possibly not as important in other company functions, can seriously affect a value engineering program. A well-managed company with strong leadership, high motivation, and an overall climate of progress and growth will provide an environment in which a value engineering program can be introduced successfully and made to flourish. Value engineering's success depends heavily on people, on their acceptance and cooperation; of especial importance is the attitude of the manager to whom the value engineering group reports. People, attitudes, and climate are particularly critical to a successful value engineering activity and must be given proper weight in designing a program.

## 10.4     Indoctrination of Top and Middle Management

Although value engineering may have the enthusiastic backing of a few key people in an organization, it will not function fully productively if it is not sold to all key members of the firm. This is particularly true for the top and middle management groups; they can either cause the value engineering program to flourish or can seal its doom by indifference and lack of support. For this reason it is strongly suggested that, before formally announcing the program, all top and middle management be provided with orientation sessions. These would brief them on the proposed program: what value engineering is all about, how it will affect them, and what their responsibilities will be with regard to the program. In addition, it would be advantageous to expose this group to a short training-type session and walk them through a typical value engineering case study. This course of action will provide them with a basic understanding of what is about to happen when the program is initiated and thus pave the way for the necessary acceptance and support of the value enginnering proposals that will eventually affect the various operating groups they manage.

Past experience shows that it is principally the middle management group that ultimately controls the destiny of value engineering programs. They are the group that must provide the supporting personnel and data required by the value engineering group; approve or disapprove value engineering proposals; and, in general, cooperate with all aims and activities of the value engineering program. Fear—created by a lack of knowledge and poor communications—can turn many middle managers against value engineering before it gets started. Elimination of this fear and creation of a proper climate requires that these initial orientation sessions be held early in the program and include all key personnel.

## 10.5    The Value Engineer or Manager

The person selected to head the value engineering group will have a decisive influence on its success or failure, probably to a greater degree than management itself. The value engineer is the key man in the whole enterprise, and to some extent the whole program stands or falls as a result of his actions. Is it possible to decide what kind of person a value engineer or manager should be?

Some necessary qualities stand out. He must be able to get along with people—be a "good mixer"—yet he must have a strong character. He must often persuade men senior to himself to act contrary to their own ideas. He must be open-minded and receptive to all new ideas, yet he must avoid the temptation to push new things at all costs. He will often have to implant ideas in other peoples' minds—ideas which they will later believe are their own. He will have to smooth the ruffled feathers of a designer who has suddenly been told that 40 percent has been cut from the cost of his brainchild while improving its performance.

Perhaps the most important characteristic is the one that all real scientists have; the value engineer must still have the naivete of a child. A scientist sometimes makes discoveries because he asks "obvious questions"; he is not afraid of asking "Why?" The value engineer must ask both "Why?" and "Why not?" at the right moments.

Some other qualities are not so important, and sometimes usually desirable qualities may even be positive disadvantages: technical knowledge, for instance. Clearly the value engineer must be able to "speak the language" of the field in which he is working, but a too specialized knowledge may put him into a ditch from which he cannot see the total landscape. Creativity and inventiveness, surprisingly, may be disadvantages. The value engineer's job is to encourage *others* to be creative. Even if he finds a clever solution, he may keep it quiet for a while in the hope that one of the team will come up with it.

He must be completely logical and capable of explaining his thought processes; he must be able to communicate clearly and unambiguously; he must have complete and obvious integrity; he must have a high standard of general intelligence. In particular, the ideal value engineer must possess social sensitivity. He must be able to persuade people whose resistance to change may be due to stubbornness, fear, or simple dislike for any changes, yet who will rationalize resistance to them by finding ingenious reasons why even obviously desirable ones cannot be effected. In short, the value engineer or manager must understand the motivations of men and be able to manipulate them for the good of a project.

It is usually the task of the value engineer or manager to select, or assist in the selection of, the value team or task force. This task should not be taken lightly since the operation's chances of success depend on a good team. Sometimes men ought to be selected who have high organizational positions even though they are not advocates of procedural change. Disastrous as this may seem, it must be remembered that the proposals of the team must eventually be implemented. The higher the position of the team members in the company, the more certain will be the implementation of the team's ideas. And it may be better to have somewhat poorer ideas implemented if the alternative is brilliant proposals that are not. The selection must therefore be based on management support and organization climate.

If it is company policy to recruit task forces regularly, it is part of the value engineer's job to see that successful members are reselected and failures quietly dropped (one must be sure, of course, that they really are failures and not slow developers.) A team should not be reselected time after time, however, no matter how good it may be; such a practice leads to laxness and sterility of ideas.

A subtle point in team selection is the relations between members. It is not always good policy to pick members on the basis of their getting on well together; mutual conflict will often spark ideas. Team conflict may make the value engineer's job tiring and frustrating, but it may also deliver the goods in ways that amity cannot.

## 10.6    Professional Background of the Value Engineer

In order to avoid jeopardizing the success of a value engineering program, it is preferable to hire an outside professional value engineer with the desired qualifications—despite the initial disadvantages arising from his lack of knowledge of company operations, personnel, and products—if there is not a suitable, competent individual available in the organization. But, since value engineering is a comparatively new profession, it is always difficult and often impossible to secure a *competent* professional value engineer. Many successful value engineering programs have been initiated and carried out completely by internal staff.

Should the management decide to initiate a value program without outside help, the value engineering manager selected must be the type of individual already described. Some tradeoffs will be necessary, of course, since no one individual will have all the personality traits, and the abilities, experience, and training desired.

The vast majority of those engaged in the field have an engineering, manufacturing, or purchasing background and are either graduate engineers or are technically competent through experience. This background is highly desirable, since improvements in product design frequently involve engineering changes that must be "sold" to engineering personnel. Getting an engineer to change his design requires—in addition to the utmost tact and resourcefulness—the ability to "speak his language."

In terms of work history, a minimum in most cases of at least five years' industrial experience is desirable. Such experience would consist, preferably, of some time spent in the areas of product design, manufacturing methods, purchasing, sales or contract administration, as well as some cost reduction work. This broad experience is desirable because the value engineering manager will be dealing with *all* company functions and should be aware of the methods of operation of each as well as the problems faced.

Finally, the value engineering manager should be, or should become, thoroughly familiar with all the techniques and methods of value engineering. This education can be obtained from training seminars offered by various educational associations and consultants and by some colleges and universities. The value engineering manager must be able to teach value engineering as well, both to his peers and his subordinates, and to instruct management through orientation sessions. To do so, he must therefore become a conscientious student of value engineering, learning all there is to know about the subject, with a level of ability respected throughout the company. Since training is a most important phase of a successful program, the value engineering manager must be able to train personnel or must seek outside assistance.

The essential basic training for a value engineering manager consists of a minimum of 30 to 40 hours of formalized instruction, followed by additional experience gained through actual work. This combination is necessary to develop sufficient expertise in value techniques and a minimum level of specialized knowledge. It is generally agreed that a training period of approximately one year is necessary to develop a competent value engineer. But this is not to say that the value program must be delayed until the manager is fully trained. The program and the training of personnel can go on at the same time, especially if consultants are used. However, the value manager should strive to receive at least some training before the program is introduced.

## 10.7    Use of Consultants

Because value engineering is a comparatively new discipline, and because there may be gaps in the capabilities of companies starting a value engineering activity, some consideration should be given to using a consultant to assist in formulating and introducing the program. There are a number of experienced and capable consultants in the United States and abroad. The advantages of using consultants are many, and their services can be tailored to fit the needs of the individual company. Their personal background, training, and experience in organizing value engineering programs in other companies can be of significant value. They can offer advice and direction on the strengths and weaknesses of various organizational arrangements, operating procedures, and personnel, as well as indoctrinate and train personnel, while counselling on pitfalls to avoid. They have the advantage of the respect usually accorded to consultants and can speak freely to both management and operating personnel. The value of advice from a third party with a minimum of personal bias is not to be overlooked.

The variety of services offered by consultants ranges from complete programs to partial ones. A complete program would include:

1. Indoctrinating top and middle management
2. Training all decision-making personnel in value engineering techniques
3. Assistance in selecting personnel for value engineering responsibilities
4. Recommending organizational arrangements as well as policies and procedures
5. Providing training and promotional and teaching aids
6. Conducting periodic program follow-up and audit.

Since companies vary in their needs for a consultant's services, the package can be varied accordingly. Perhaps a company may wish to do no more at first than train several of its personnel in value engineering techniques. This aim can be achieved by sending personnel to a workshop seminar conducted by one of those consultants who periodically appear in different cities throughout the country.

Whatever decision a company makes regarding the use of consultants, that decision should be made with a full knowledge of the services they offer.

## 10.8    Program Initiation

Having gained some knowledge of value engineering and reviewed the various types of organizational structures suitable for this activity, as well as the many factors that affect it, a company can now begin to design a value program. The first order of business is to answer these questions about the activity:

1. Where will it be placed in the organization?
2. To whom will it report?
3. Who will head it up?
4. Will it be a staff or line function?
5. Will it be centralized or decentralized?
6. How many members will be needed?
7. How will it function?

The total description of the value program should be outlined in a corporate policy statement, to be issued at the initiation of the program. This policy statement will answer the questions above, as well as clarify such things as (1) management goals and objectives in instituting the program, (2) management support and endorsement of the program, (3) the duties of both value and engineering personnel, (4) the operating functions affected by value engineering, and (5) the operating and reporting procedures required. An example of a typical policy statement is included in Appendix A.

By issuing such a statement, management can put to rest many of the doubts and misgivings that are bound to arise at the introduction of something new in the company and prevent its becoming suspect, at least to department heads and managers. In this as in all human relations, nothing breeds fear and distrust like ignorance and misunderstanding. The person who feels threatened by any new activity will certainly not cooperate with it and will more than likely oppose it either passively or actively. This difficulty can be minimized by giving department heads and managers a clear and well-defined explanation of all facets of the program. Effort devoted to maximizing communication will be amply rewarded. Finally, the benefits to be derived from judicious publicity must not be overlooked. Such things as news releases, write-ups in local papers and house organs, and announcements on bulletin boards can be used effectively to state program objectives and to promote support and participation.

## 10.9    Formal Training Programs

One of the first objectives following the introduction of a value program should be to train all decision-making personnel in value engineering techniques; in this way the benefits can be reaped as soon as possible. Decision-making personnel, once trained, can begin to apply these techniques to their own areas of responsibility and encourage others to do the same. Ideally, if all personnel used value engineering techniques in solving problems and making decisions—particularly those that affect cost—the maximum benefits of a value program would be achieved.

A second benefit of no less importance is the broad

base of support that training establishes for the value program. People who have been fully briefed about value engineering techniques understand the nature and objectives of the program and are willing to support and participate in its efforts. This support might take the form of supplying a value engineer with all the data needed for a project or of merely sitting in as member of a task force study team to improve the value of a product. Whatever the situation, a person's previous exposure to a training seminar does much to improve the quality and effectiveness of his contribution to a project. It has also been found that a person who has successfully applied value engineering techniques usually becomes a staunch advocate of the value program and will do much to further its goals.

Various degrees of training can be provided, depending on the company's aim in training any one group. The shortest profitable exposure is about two hours (usually referred to as an orientation session); it is used to indoctrinate top and middle management in the general principles of value engineering.

The next level of training is a one-day orientation and workshop seminar that is used to acquaint middle managers and department heads with the working principles of value engineering. The day is usually divided in half, the first session consisting of background material and exposure to value engineering techniques and fundamentals, the second devoted to their application to an actual project. Because of the time limitation, the instructor usually guides the group through a simple project permitting the fullest application possible.

This degree of training is limited in its value because the participant does not have enough time either to develop facility in the techniques or to fully appreciate their potential. It can also leave the participant with the impression of oversimplification. However, it does provide at least limited exposure for those who are too busy, or are unwilling, to take part in a full workshop seminar.

The third level of training is the 30- to 40-hour workshop seminar, which is usually conducted over a one- or two-week period (though it can be extended over a number of weeks). This type of training is a must for all decision-making personnel at the working level. It includes all engineers, designers, production and manufacturing personnel, buyers, and quality control personnel. In addition, training should be extended to include personnel from accounting, sales, and contracts. The full workshop seminar is the most frequently employed training method and has been found to be the most effective; it may be extended to 60 or 80 hours to introduce more sophisticated techniques.

Training is of two general types: exposure to value engineering techniques and fundamentals and the application of these techniques and fundamentals to hardware projects. Giving the participant sufficient in-depth instruction in value engineering and then having him apply his knowledge to an actual project allows him to experience its effectiveness in reducing costs and improving value. He is then more inclined to utilize these techniques when he returns to his regular assignment; this, of course, is the major objective in having him trained.

It is naturally far more desirable to provide good training facilities than to depend on some makeshift arrangement. If the company has no formal training rooms, holding the workshop away from the plant has definite advantages. The proper atmosphere for learning can be created more easily under such conditions for there are very few business interruptions, people are more relaxed and receptive, and a certain status will be conferred on those selected to attend.

The use of consultants was discussed in an earlier section. It should be mentioned here that it is precisely in the area of training that they have the most to offer. Since the major part of a consultant's contribution in helping establish a value program consists in training personnel, this is the area in which most of them are most talented. Doing such work continually has given them a certain expertise which the amateur cannot match. Since a good training program is critical to a good value program, the use of a reputable consultant is strongly recommended, particularly if the "in-house" capability is weak.

Typical formats and agenda for training seminars can be found in Appendix B.

## 10.10 Selection of Personnel

The premature selection of unsuitable individuals can cause awkward and embarrassing situations if they must subsequently be dismissed; such situations can only hurt the value program. Many companies defer the selection of people for value work until after the training program, thus allowing management to study those attending the workshop and to pinpoint the individuals who exhibit the greatest talent. If a consultant is used for the training function, he can be unusally helpful in suggesting individuals whom he recognizes as outstanding. His observations, coupled with those of management, will help to insure judicious selections. A workshop will often produce individuals with a marked talent for and interest in value work not previously suspected. It is exactly this combination of talent and interest that is looked for during the seminar.

As an additional aid in helping to select persons for value work, a position description is included in Appendix C.

## 10.11    Program Operation

Once a value program has been initiated and personnel responsibilities assigned, a new organizational function exists that must mesh with the overall operation of the business as smoothly and effectively as possible. The responsibilities of the value group as outlined in the corporate policy statement cannot be fulfilled until specific procedures to govern the group's operation are developed. Working relationships must be set up with other company functions and overlapping areas of responsibilities clarified. Criteria must be established for selecting value projects, for evaluating alternatives, and for deciding on the merits of various suggestions. These and many other details must be worked out by the value group before an efficient and effective value program can truly get under way.

Time and effort must be devoted at the very beginning to the development of a clear, well-defined understanding of the value activity and its relationship to other company functions so as to eliminate mistrust and possible friction later in the program. The development of sensible procedures and operating practices requires a thorough knowledge of the value group's responsibilities and goals. With this in mind, the group's major responsibilities have been divided into the following broad categories:

1. Informal training and education
2. Product evaluation
3. Consultation
4. Government and contractor liaison
5. Value standard and data development
6. Calculation of savings and reporting.

It is true that some changes in procedures will be

made as the value program develops, but if a concerted effort is made in the beginning, future changes will be minimal.

## 10.12    Informal Follow-Up Training

Beyond the formal training programs whose strengths and weaknesses have been discussed, there is a need for follow-up training if value techniques are to be firmly ingrained. Without follow-up training those who have completed formal training programs are liable to slip back into their old habits and fail to utilize fully the techniques they have learned.

Basically, follow-up training provides reinforcement of what has been learned in the formal training program. Reinforcement is a training procedure to which nearly everyone has been exposed at home, at school, or at work. It is particularly beneficial in value engineering training, and there are several methods of achieving it.

One effective method of reinforcement is to use personnel in what is called a task force group. People are temporarily assigned to a task force group for a specific value project. (The use of such a group is usually warranted when improving the value of some product or system will be of considerably greater significance if accomplished in a short time.) The task force group may work from several hours per week to full time, depending on circumstances.

On the task force project the opportunities to reinforce value training are many, and the value engineer should be constantly ready to take advantage of them. Since the tendency to use shortcuts can be strong, especially if the pressure to accomplish is high, it is the value engineer's job to see that any halfway approach is avoided. He must make sure that the group proves to itself the merits of full utilization of all required techniques. The surprising accomplishments of a group run in this way will strongly influence each member to apply these techniques when he returns to his regular assignment and to encourage others to do the same.

Another opportunity for follow-up training is attendance at design reviews. By properly preparing for such sessions, the value engineer can frequently offer superior solutions to design problems, challenge high cost areas, stimulate the group to better design practices, and, in general, promote greater use of value techniques in solving design problems. A short but typical checklist of questions useful in promoting better value under such circumstances is included in Appendix D.

A word of caution: Since a designer is naturally going to be partial to his own work, the value engineer must avoid situations that would embarrass him. If changes for value improvement cannot be presented without subjecting the existing design and its designer to excessive criticism, it would be wiser for the value engineer to make his suggestions privately to the designer. This move will make it more likely that value engineering will gain a supporter rather than an enemy, and the project at hand will also benefit.

## 10.13    Advanced Training

A discussion of education and training would be incomplete without touching on advanced training for value engineers. It is essential that value engineers receive advanced training beyond the basic workshop course. Participation in workshops as an instructor is a good means of developing a fuller understanding and mastery of value techniques. Attendance, either as a student or as an instructor, at workshops inaugurated by other companies or training facilities also helps to broaden the value engineer's background by exposing him to some variations of value techniques used by other groups.

At various times throughout the year, opportunities are offered by educational institutions for advanced training in subjects dealing with more sophisticated value engineering concepts such as (1) the use of PERT/COST in value projects, (2) the development and use of learning curves, (3) the development of value standards and the theoretical evaluation of functions, (4) the use of creative techniques both individually and with groups, (5) the management and control of value programs, and (6) government requirements and relationships regarding value engineering. Such training is sometimes offered under the auspices of the Society of American Value Engineers (SAVE), in conjunction with a college or university. Members of the society are usually informed of these activities through their society bulletin; a monthly newsletter, published by an independent Washington, D.C. firm, contains a listing of all such events.

Value engineers should also attend industrial and professional shows and exhibits. As in all professional groups, there is no substitute for keeping abreast of the latest developments in one's chosen field.

## 10.14    Product Evaluation

The major portion of a value engineer's time and energy will most likely be spent in product evaluation. This involves the selection and study of products, systems or services, whether new or old, that offer the most potential for value improvement. It is usually the value group's responsibility to base the project selections on the evaluation of a number of factors, all of which have a dollars-and-cents bearing on the selection. Occasionally, projects are requested by management for specific reasons, such as poor markup or stiff competition, but this is the exception rather than the rule.

Since the overriding objective of the value program is to improve value and thereby improve profits, it stands to reason that products or services that involve the greatest expenditure of the company's cash should be first on the list for review. Whether these are high-or low-volume, expensive or inexpensive items, or standard components used in a wide variety of end products, the total expenditure will indicate the prime candidates for a value study.

It sometimes happens that a value study is warranted on a new, fast-growing product because of its obvious need for improvement and its strong sales potential despite the fact that sales have not yet reached substantial volume. A product with a declining sales curve is probably not a good candidate unless it is felt that cost improvement could make it competitive enough to increase sales and improve profits.

On occasion, requirements of the government or prime contractors may dictate that a value study be made on a given product. In such cases the potential for additional profits should be fully explored. It has frequently happened that companies have made larger profits through value incentive clauses in contracts than through the profit margins allowed in those contracts.

After projects are selected for study, they can be assigned to individual value engineers, the value department as a group, or to a task force team, depending on the scope of the project. Whatever procedure is followed, it is the duty of one member of the value group to summarize and present all suggestions developed by the study for value improvement to the responsible authority. He would include all of the pertinent technical and economic data, and as much other information as is required by that authority to make a proper evaluation and decision. Incomplete and poorly prepared value presentations will receive cursory treatment by the reviewing authority. The time, effort, and money expended in making a value study should not be neutralized by a poor presentation.

Since it frequently happens that value suggestions do not demand immediate action, the reviewing authority can push them aside in favor of more pressing every-day problems. Under such circumstances the responsible value engineer must give way tactfully and seek to have the proposals implemented in some other fashion if unacceptable in their original form. Follow-up is sometimes the most difficult and frustrating phase of a value study, but until the program is implemented, value improvement does not become a reality.

## 10.15    Within-Company Consultation

As the value engineering activity progresses in a company, the value group will periodically be requested by various operating sections to assist in a search for better value. Such assistance might take the form of conducting value studies; making product and process searches; estimating costs of various design alternatives; assisting in setting cost targets; assisting vendors in reducing costs and prices; assisting purchasing in selecting functionally equivalent products; and so on. In such situations value personnel act as consultants and seek to bring their special value training to bear on the problems presented.

If the value engineer is invited by another operating section to assist in improving value, he may assume that a major roadblock has been bypassed; obviously the section is already questioning existing conditions and is willing to change if change offers hope of improvement. Also, by initiating and being part of the study the section that made the request has taken a proprietary interest in improvement and will frequently devote considerable effort to it.

When the VE activity is requested to provide its services, its primary function is to develop value alternatives for the requesting party, not to attempt to make decisions or control implementation. Such behavior could easily alienate the requesting party and destroy the climate for improvement. On the other hand, a successful encounter with VE will generate enthusiasm for the use of value engineering techniques on other projects.

## 10.16    Value Standards and Data Development

Value decisions depend on having accurate cost information, and the value group is responsible for providing as much assistance in this area as possible. Information on the cost of materials, processes, basic machining and production operations, and the like, should be made available to engineering and design personnel to the fullest possible extent. When pressure is on the design group to design products to meet rigid specifications and to be available in a short period of time, the cost parameter is (understandably) frequently neglected. However, if meaningful cost information is available at the time that design decisions must be made, the achievement of value goals may still be possible. The value group should always develop data on cost of functions that can be useful at the design phase. Such information can provide design direction early in the design phase as well as assist in establishing cost targets.

## 10.17    Government Contracts

Working on government contracts, either directly or through a prime contractor, sometimes offers special opportunities for profit improvement through value engineering. Most contracts over a certain minimum price include value engineering incentive clauses. If a contract does not contain such a clause, it can usually be amended to include one at the request of the contractor. These incentive clauses are not mandatory, and the contractor may choose to ignore them. If he does so without a thorough review of the contract, however, it can mean lost dollars.

The purpose of the value incentive clause is to motivate government contractors to reduce the cost of contract performance by providing for sharing the money saved as a result of cost saving suggestions initiated by the contractor. Since the various contract types and sharing arrangements are discussed in detail in Section 17 of the Armed Services' Procurement Regulations, only a brief description of the usual procedures is in order here.

Upon receipt of a contract containing a value engineering incentive clause and, after studying ways that contract performance costs can be reduced, the contractor prepares Value Engineering Change Proposals (V.E.C.P.'s) to be submitted to the government or to the prime contractor. If his proposals are accepted by the government, the contract is amended to include the suggestions, and the contractor and government share the savings based on a formula included in the contract. This description is an oversimplification of the entire procedure but does give the gist of it.

With another type of value engineering contract arrangement called a value engineering program clause, the government funds the expense of the contractor's value engineering effort, but the contractor's percentage of the savings is substantially reduced. This plan is ordinarily used for research and development type contracts in which the pinpointing and calculation of savings are difficult.

The responsibility of the value engineering activity in dealing with such contracts is to become fully informed regarding the details of operations, to obtain the most attractive terms for itself, and to be safeguarded against the losses that can occur from carelessness.

For all the roadblocks and the red tape encountered by firms in dealing with the government in this area, V.E.C.P.'s have earned substantial additional profits for companies time and again. Moreover, successful value engineering programs under these federally weighted guidelines for determining contractor performance and profit allowances have not only helped companies secure new government contracts but also increased their total profit potential.

## 10.18     Calculating Savings and Reporting

Like other company activities, value engineering must report on its progress regularly, indicating current project status and future goals. On the basis of such reports, management can determine whether the activity is operating satisfactorily, whether it should be expanded or curtailed, and whether any changes are necessary. Such reports also provide the value engineering activity with an incentive for accomplishment and a chance to inform management of the benefits being derived from its efforts (provided, of course, that the program is successful).

There are several criteria for measuring a program's success; the most common one simply compares the amount of money saved to the cost of the program. Calculating savings may sound rather simple, and so it is when a product or service already exists. In that case, the change in cost or performance is readily ascertained. However, since value engineering is employed early in the design cycle to prevent unnecessary costs from ever getting into a design, it becomes more difficult to measure savings. It is possible to estimate them under these conditions only if design alternatives have been formalized to the degree that permits it; otherwise it is foolish to attempt to evaluate costs and savings. Management must use other means of measurement, such as the quality and simplicity of the final design, the competitive status of the end product, the profitability of the product, its ease of manufacture, and its performance.

In measuring value savings, most companies prefer to think in terms of the savings that will accrue during the first year of the change. Although this is an arbitrary time period, it is the one in common use. If changes are made on products for which contracts have been arranged for periods longer than one year, the total contract time is sometimes used in calculating savings. Such a procedure is obviously valid, provided the contracts are firm.

In selecting the elements of product cost to be used in determining savings, many companies prefer to use what is referred to as "out-of-pocket" costs. These costs usually represent the material, direct labor, and variable overhead portions of the total cost of the product and are directly affected by changes in the product. To include *all* overhead and loading charges when comparing the new and old costs will tend to understate savings. If specific fixed overhead items have actually been reduced by incorporating the changes, these reductions should be reported. In addition, all nonrecurring costs of implementation should be deducted from the proposed savings so that a maximum return-on-investment ratio is presented to management. If the value group is competent and conscientious, it will have no difficulty finding profitable projects using this basis for calculating savings. The reader is referred to Appendix E for an example of tally sheets used to calculate savings on such a basis.

Costing methods and procedures should be worked out between the value group and the accounting department, with the latter being responsible for the auditing of all reported savings. This arrangement will insure accuracy and upgrade the confidence level of savings reports. In government contracts the procedure has been made almost mandatory by government regulations governing value engineering.

A ratio of dollars saved to cost of program (5:1 or 10:1 is most common) can be used as a guideline in judging the results of programs in different divisions within a single company, but keep in mind when comparing these results to those obtained by other companies that accounting procedures vary and that only comparisons between companies using the same procedures are valid.

Frequency of reporting is flexible. Some companies use a monthly, others a quarterly reporting system. Each management must decide for itself the frequency necessary for proper evaluation and control. A monthly system may be advisable initially, with a change to a quarterly system after a program has been functioning effectively for some period of time.

Another criterion frequently made use of in value engineering reports is the overall worth of the proposals made by the value group based on the number accepted and the number rejected. This factor reflects both on the quality of the proposals and on the value group's success in having them implemented. A large number of rejected proposals is an indication of trouble and a need for management action.

As mentioned earlier, although the savings resulting from value engineering efforts early in the design phase are frequently difficult or impossible to measure, they are nevertheless quite real. An indication of how much of this type of work has been done should be included in value engineering reports by reporting hours devoted to consultation, design reviews, investigations, and the number of requests for value services by other functional groups in the company.

Accomplishments in the area of training should also be reported, since this is a prime responsibility of the value group. Such training efforts have their own decided effect on savings (though these savings can be measured only in the broadest terms of company growth and accomplishment). Trained personnel using value engineering techniques in their daily operations can produce considerable savings for a company.

The objective of this chapter has been to present to the reader all the information necessary to design an effective value program to fit the needs of his particular company. It is hoped that the discussions of the various phases of such a program fulfill this purpose.

# REFERENCES

1. Miles, Lawrence D., *Techniques of Value Analysis and Engineering*, McGraw-Hill, 1961.
2. *A Guide for the Organization and Operation of Value Engineering*, Electronic Industries Association, Value Engineering Subcommittee, Washington, D.C.
3. *Handbook H-111 on Value Engineering* and *Principles and Applications of Value Engineering*, U.S. Department of Defense, Washington, D.C.

# APPENDIX A

## TYPICAL POLICY STATEMENT OF CORPORATE PROCEDURES

I. PURPOSE

To establish an effective value engineering policy for the — — Company and set forth basic requirements for its implementation.

II. DEFINITION

Value engineering is the term applied to all actions that identify and eliminate unnecessary cost in design, development, procurement, manufacture and delivery of a product or service, without sacrificing essential quality, performance, reliability, or maintainability. It is a functionally oriented and planned effort to attain the optimum relationship of performance, reliability, and cost.

III. POLICY

It shall be the policy of the — — Company to fully and effectively utilize the principles of value engineering in all areas of operation. Particular emphasis is to be placed on applying value engineering to product design as early as possible as well as throughout all stages in the design, development, and manufacturing cycle.

IV. RESPONSIBILITY

The Director of Engineering and Research shall retain overall responsibility for the value engineering program but shall delegate its administration to the Manager of Value Engineering. Similarly, the responsible director of each operating group shall be responsible for value engineering activities within his group but shall delegate functional control to the Manager of Value Engineering. Specific responsibilities of the value engineering manager are the following:

    1. To organize and implement a value engineering program throughout the company.

    2. To develop and establish value engineering policy and objectives.

3. To provide education and training of company personnel in value engineering principles and techniques.

4. To develop value standards and cost data for use of operating personnel.

5. To develop operating procedures for implementing and monitoring the value engineering function in all company operations

6. To supervise and direct the value engineering function in the various operating groups.

7. To set up and direct value engineering task force groups to work on special projects.

8. To report to management on the progress of the program.

9. To take responsibility for value engineering contractual commitments with the U.S. government or governmental contractors.

10. To represent the company in external activities dealing with value engineering.

## V. ORGANIZATION

The Value Engineering Manager shall report to the Director of Engineering and Research. He shall supervise and direct the value engineering function at corporation headquarters both administratively and functionally and shall exercise functional control over this activity at company facilities at other locations.

## VI. PROCEDURE

### 1. Initiation of Projects

Projects will usually be initiated by the value engineering function but may be initiated by management, purchasing, or other sources within the company. A preliminary review will be made by the value engineer to determine savings potential, scope of work, and probability of implementation. Projects with potential savings will be assigned a project number and become the value engineer's responsibility to follow through to a conclusion.

### 2. Method of Operation

Depending on the size of a project, it will be handled by the value engineer individually, or a task force effort may be required. If the latter is necessary, the Value Engineer Manager will establish with the value engineer a plan for the project which will involve the following:

a. Determining the scope of the project.

b. Selecting required number of task force teams, arranging for temporary release of personnel assigned to teams and allocation of costs.

c. Accumulation of necessary data for project work.

d. Scheduling and supervising of task force effort including establishment of cost targets.

e. Arranging for necessary customer and vendor liaison.

f. Accumulation of task force results and preparation of final proposals.

g. Submission and follow-up of proposals for implementation.

## 3. Implementation Authority

The responsible department head receiving a value engineering proposal retains the autority to accept or reject such a proposal. However, if he rejects or otherwise fails to implement the proposal, he must provide a written explanation for his decision. If a dispute should arise regarding the implementation of a proposal, final authority will rest with a review board consisting of the Director of Engineering and Research, the Sales Manager, the Production Manager, and the Manager of Value Engineering.

## VII. REPORTING

The Value Engineering Manager will submit a quarterly composite report to Corporation management summarizing the results of the value engineering function in the various operating groups in the company.

# APPENDIX B

## TYPICAL VALUE ENGINEERING WORKSHOP SEMINAR CURRICULUM

| | |
|---|---|
| 10 minutes | Keynote |
| 20 minutes | Value Engineering's History, Concepts, Philosophy |
| 20 minutes | General Orientation of Value Engineering Techniques |
| 20 minutes | The Importance of Evaluating our Habits and Attitudes |
| 10 minutes | Recognition of "Roadblocks" and Overcoming Tehm |
| 15 minutes | What Value Engineering Can do for this Division or Operation |
| 10 minutes | Break |
| 15 minutes | Selection of Product for Study |
| 15 minutes | Get All the Facts |
| 15 minutes | Determine Costs |
| 30 minutes | Determine the Function |
| 30 minutes | Functional Workshop |
| 1 hour, 15 min. | Lunch |
| 15 minutes | Put a $ on the Specifications and Requirements |
| 30 minutes | Functional Workshop |
| 3 hours, 30 min. | Project Work (Gather Project Information) |

SESSION II

Project Work (Gather Project Information, Correlate Project In-Information, and Determine Project Function)

SESSION III

| | |
|---|---|
| 15 minutes | Developing Alternatives |
| 30 minutes | Creativity |
| 30 minutes | Creative Workshop |

| | |
|---|---|
| 15 minutes | Blast and Create |
| 15 minutes | Break |
| 2 hours | Creative Workshop (on Operation Projects) |
| 1 hour, 15 min. | Lunch |
| 1 hour, 45 min. | Project Work (Creative Session on Project) |
| 2 hours | Determine Function and Create |

SESSION IV

| | |
|---|---|
| 30 minutes | Recap on Information Gathering and Development of Alternative Phases |
| 3 hours, 15 min. | Project Work (Correlate Information From Functional and Creative Efforts) |
| 1 hour, 15 min. | Lunch |

SESSION V

| | |
|---|---|
| 30 minutes | Every Idea Can Be Developed |
| 15 minutes | The Importance of Testing and Verification |
| 15 minutes | How To Refine Ideas |
| 15 minutes | Put a $ on Each Idea |
| 15 minutes | Evaluate the Function |
| 15 minutes | Evaluate by Comparison |
| 15 minutes | Break |
| 1 hour, 45 min. | Project Work (Evaluate Ideas) |
| 1 hour, 15 min. | Lunch |
| 15 minutes | The Use of Specialty Vendors |
| 15 minutes | Consult Vendors |
| 15 minutes | Use Specialty Products, Processes and Materials |
| 15 minutes | Use Company and Industrial Specialists |
| 15 minutes | Use Company and Industrial Standards |
| 15 minutes | Break |
| 2 hours, 15 min. | Project Work (Investigation of Project Ideas) |

SESSION VI

| | |
|---|---|
| 60 minutes | Introduction of Specialty Suppliers |
| 15 minutes | Vendor Display Time |
| 15 minutes | Break |
| 2 hours | Project Work |
| 1 hour, 15 min. | Lunch |

| | |
|---|---|
| 60 minutes | Introduction of Specialty Suppliers |
| 15 minutes | Vendor Display Time |
| 15 minutes | Break |
| 2 hours, 15 min. | Project Work |

SESSION VII

Project Work

SESSION VIII

Project Work

SESSION IX

| | |
|---|---|
| 3 hours, 45 min. | Project Work |
| 1 hour, 15 min. | Lunch |
| 15 minutes | Developing the Proposal |
| 15 minutes | Motivate Positive Action |

SESSION X

| | |
|---|---|
| 3 hours, 45 min. | Project Work |
| 1 hour, 15 min. | Lunch |
| 30 minutes | Value Engineering—a New Tool for Everyone To Use |
| 1 hour, 45 min. | Project Work (Wrap-up) |
| | Management Presentation |

# APPENDIX C

## TYPICAL JOB DESCRIPTION FOR A VALUE ENGINEER

I. JOB SUMMARY

The value engineer has the broad responsibility:

1. To insure that value engineering techniques are properly applied to the projects assigned so that such projects are accomplished at minimum costs.

2. To provide value engineering support to all areas, particularly engineering, design, purchasing and manufacturing.

3. To promote the use of value techniques through training activities and personal contact with company personnel.

II. DUTIES AND RESPONSIBILITIES

*Major Duties and Responsibilities*

1. Select product areas that will yield the most profitable results from the application of value engineering techniques.

2. Conduct group value engineering workshops on specific products and designs.

3. Apply value engineering techniques at various phases of the product cycle, from specification and design reviews through procurement and manufacture.

4. Develop value proposals, submit them to the responsible department manager, and follow up their implementation.

5. Provide data on costs, methods, processes, materials, vendors and specialty products.

6. Locate and use the talents of specialty suppliers.

7. Assist buyers in reducing cost of purchased materials.

*Minor Duties and Responsibilities*

1. Stimulate line people to maximum effort in cost prevention and cost reduction through the use of value engineering techniques.

2. Assist the Value Engineering Manager on special projects and in arranging and conducting training seminars and workshops.

3. Search for materials, methods or processes to improve the value of company products.

4. Report to the Value Engineering Manager monthly on project status.

5. Assist the Value Engineering Manager in developing new value techniques to reduce costs.

### III. EDUCATION REQUIRED

College degree in engineering or science preferred, plus minimum of 40 hours of workshop training in value engineering.

### IV. PERSONALITY TRAITS

The value engineer must be creative, articulate, easily accepted by others, persevering and dedicated; he must have the initiative to undertake difficult tasks and not be discouraged by roadblocks or failures; he must be mature in thought and action.

### V. EXPERIENCE

The value engineer should have at least ten years of broad experience. Preferably, 2 to 3 years in manufacturing; 2 to 3 years in liaison activity or sales; 2 to 3 years in electromechanical design; one year in purchasing or working with vendors; one year in cost reduction or value engineering.

### VI. ON-THE-JOB TRAINING

Training period should be a minimum of six months (but preferably one year).

### VII. CONTACTS

Position involves high degree of contact with people throughout the entire company plus outside contacts with vendors and customers.

### VIII. SUPERVISION RECEIVED

The value engineer is under the direction of the Manager of Value Engineering, but the nature of his job is such that he must be capable of good performance with a minimum of supervision.

### IX. SUPERVISION EXERCISED

Since the value engineer must temporarily supervise people assigned to specific projects, some supervisory experience is desirable.

# APPENDIX D

## TYPICAL VA/VE DESIGN REVIEW CHECK LIST

1. Have the specifications been reviewed to determine if they are excessive in any way?     Yes__ No__

2. Have contractual requirements been reviewed for unnecessary costs?     Yes__ No__

3. Have potential cost trade-offs been reviewed with the customer?     Yes__ No__

4. Have the cost targets established for specific functional design areas been met?     Yes__ No__

5. Have commercially available assemblies and components been compared to find those that are most technically acceptable and economically attractive?     Yes__ No__

6. Have nonfunctional requirements been challenged?     Yes__ No__

7. Have total product life maintenance and service costs been considered in selecting specific design alternatives?     Yes__ No__

8. Have suppliers quoted on the basis of functional requirements rather than specific designs?     Yes__ No__

9. Have sufficient design alternatives been proposed and evaluated to make a proper selection of the design of assemblies and subassemblies?     Yes__ No__

10. Have materials and finishing requirements been reviewed for lower-cost alternatives?     Yes__ No__

1. Have alternative manufacturing processes been considered in designing components?     Yes__ No__

12. Have material removal costs been minimized?     Yes__ No__

13. Have minimum tolerances been specified where practical?     Yes__ No__

14. Have assembly inspection and testing costs been considered in the designs?     Yes__ No__

15. Have packaging, shipping, and storage costs been considered in evaluating design alternatives?     Yes__ No__

16. Have assembly costs been reduced by combining components?     Yes__ No__

17. Have size variations been minimized for comparable components? Yes__ No__

18. Has the design been planned to reduce machining set-ups and the need for special equipment? Yes__ No__

19. Have paperwork and overall software requirements been reviewed for the possibility of savings? Yes__ No__

NAME:_____

TITLE: _____

# APPENDIX E

## TALLY SHEETS USED TO CALCULATE SAVINGS

*VALUE ENGINEERING*
*SUMMARY*

| Part name | | | | | Part number |
|---|---|---|---|---|---|
| Where used | | | | | Annual quantity |

| | Material | Direct labor | Direct overhead | Total | Expenses |
|---|---|---|---|---|---|
| Original cost | | | | | Tooling |
| Revised cost — Est. | | | | | Design and drafting |
| Revised cost — Act. | | | | | Prototype manufacturing and test |
| Savings/piece — Est. | | | | | Approval agency fees |
| Savings/piece — Act. | | | | | Other |
| Annual savings — Est. | | | | | |
| Annual savings — Act. | | | | | TOTAL |

| | 1st Year | 2nd Year | 3rd Year | 4th Year |
|---|---|---|---|---|
| Gross savings | | | | |
| Less expenses | | | | |
| Net savings | | | | |

*Description of Suggested Change:*

Originated by:          Completion Date:

# VALUE ENGINEERING
# INFORMATION PHASE

Work Sheet No. 1

Project No.:_____

Product:_____ Part No.: _____

Review the following checklist for the information required prior to starting project.

A. *Marketing and Cost Data*

    1. Present volume and future potential at various cost levels
    2. Customer requirements and necessary features
    3. Complete cost breakdown (material, labor, and overhead)

B. *Engineering Data*

    1. Design and development history and future potential
    2. Functional requirements
    3. Specifications: Company, customer, or approval agency
    4. Drawings and material lists
    5. Set of parts or assembly

C. *Manufacturing and Purchasing Data*

    1. Operation sheets, rates, and production quantities
    2. Special tools, fixtures, and equipment required
    3. Manufacturing problem areas and scrap rate
    4. Packing and shipping requirements
    5. Vendors and quantities purchased
    6. Problem areas and recommendations

Team Members:_____ Team No.:_____ Date:_____

# VALUE ENGINEERING
# DEFINITION PHASE 1

Work Sheet No. 2

Project No.: _____

Product: _____ Part No.: _____

1. Describe all the functions of the product using a verb and a noun only.
2. Now select the function that best describes the purpose for which the product exists; in other words, its basic function. All other functions are secondary.

| Functions | | Basic or |
| --- | --- | --- |
| *Verb* | *Noun* | *secondary* |

Team Members: _____ Team No.: _____ Date: _____

# VALUE ENGINEERING
# DEFINITION PHASE 2

Work Sheet No. 3
Project No.:_____

Product:_____ Part No.: _____

1. List all component parts and the basic function of each.
2. Assign a value to each function.
3. List the cost of the components and total the "Value" and "Cost" columns. The ratio of value to cost indicates the degree of value improvement potential.

| Component part | Basic function | | Value | Cost |
| | Verb | Noun | | |
| --- | --- | --- | --- | --- |

Team Members:_____ Team No.:_____ Date:_____

# VALUE ENGINEERING
# SEARCH PHASE

Work Sheet No. 4

Project No.:_____

Product:_____ Part No.:_____

Basic Function: _____

Having defined the basic function, search for and list all ideas for new ways to provide it, using the techniques of (1) creative thinking, (2) blast and refine, and (3) comparison. List all the ideas, no matter how ridiculous, but don't evaluate them now.

*List of Ideas*

Team Members:_____ Team No.:_____ Date:_____

# VALUE ENGINEERING
# EVALUATION PHASE

Work Sheet No. 5

Project No.:_____

Product:_____ Part No.:_____

1. Review all ideas on work sheet No. 4 and place an estimated value on them in the right-hand column.
2. Select the best ideas and list them below.
3. List the advantages and disadvantages of each of them in terms of the work required to implement them.
4. Determine an order of priority and a plan of action for the ideas (or idea) selected.

*Best Ideas from Work Sheet No. 4     Advantages     Disadvantages*

*Plan of Action for Ideas:*

Team Members:_____ Team No.:_____ Date: _____

# VALUE ENGINEERING
# VENDOR QUOTATION SUMMARY

Worksheet No. 6:

Project No.: _____ —

Product: _____

Part Name: _____

Drawing No.: _____

| Vendor and process | Quotation | Tools | Delivery | Remarks |
|---|---|---|---|---|

Team Members: _____ Team No.: _____ Date: _____

# *Index*